Evolutionary Restraints

Evolutionary Restraints

The Contentious History of Group Selection

MARK E. BORRELLO

THE UNIVERSITY OF CHICAGO PRESS CHICAGO AND LONDON

The University of Chicago Press, Chicago 60637
The University of Chicago Press, Ltd., London

Paperback edition 2012
Printed in the United States of America

21 20 19 18 17 16 15 14 13 12 2 3 4 5 6

ISBN-13: 978-0-226-06701-8 (cloth)
ISBN-13: 978-0-226-06703-2 (paper)
ISBN-10: 0-226-06701-7 (cloth)
ISBN-10: 0-226-06703-3 (paper)

Library of Congress Cataloging-in-Publication Data

Borrello, Mark E.
Evolutionary restraints : the contentious history of group selection / Mark E. Borrello.
p. cm.
Includes bibliographical references and index.
ISBN-13: 978-0-226-06701-8 (cloth : alk. paper)
ISBN-10: 0-226-06701-7 (cloth : alk. paper) 1. Group selection (Evolution)—History—20th century. 2. Wynne-Edwards, Vero Copner. I. Title.
QH376.B67 2010
576.8'2—dc22

2009051465

⊗ This paper meets the requirements of ANSI/NISO Z39.48-1992 (Permanence of Paper).

Contents

Acknowledgments

I am grateful to a number of people for the contributions they have made to the successful completion of this book. I began thinking about cooperative breeding behavior in birds as a biology student at Beloit College and was fortunate to have been guided by Professors John Jungck and Ken Yasukawa to think carefully about evolutionary processes and to consider the historical and philosophical issues that were germane to a proper understanding and explanation of them. Their influence led me to pursue graduate study in the history and philosophy of science at Indiana University, where this project first took shape.

At Indiana it was Frederick B. Churchill who trained me to think and write like a historian. In our lengthy conversations after seminars, he helped me hone my ideas about group selection theory and organize them into a project that was coherent and compelling. I value immensely the time Fred took to work over my writing and to demonstrate in his own work how essential a true understanding of the scientist's ideas and context was to getting the history right. I am also deeply indebted to Elisabeth A. Lloyd. As perfect a mentor Fred was as a historian, Lisa provided the ideal philosophical foil. She sharpened my analytical abilities and helped me to wade through the mountains of philosophical literature on the levels of selection question. Lisa was an invaluable guide, and she was also incredibly generous in introducing me and my work to the philosophers who had written those mountains of literature on the levels of selection. These informal conversations with David Hull, Elliott Sober, Bill Wimsatt, and Michael Ruse, among others, convinced me that the history of group selection needed to be written and might inform the philosophical and scientific conversation.

It seems as though every successful project is the result of good fortune almost as much as it is hard work. Michael Wade joined the faculty at Indiana as I began writing. Mike is one of those rare biologists who is truly historically and philosophically inclined. He has been a consistently careful reader of this manuscript and has helped me to see the relationship between the early models of Sewall Wright and their connection to Wynne-Edwards work, among other insights.

As I conducted the research for this book I have spent time in the archives at Wynne-Edwards's home institution Aberdeen University in Aberdeen, Scotland. At Aberdeen my conversations with Alan Knox, manager of historical collections at the university archive and a PhD student of Wynne-Edwards's, were incredibly helpful. I am also grateful to the chair of the Zoology Department at Aberdeen, Regius Professor Paul Racey, for the insight he provided regarding some of Wynne-Edwards's postretirement work on group selection. At the Alexander Library in the zoology department at Oxford University, I am particularly thankful to Mike Wilson for providing access to and guidance in the David Lack papers. Finally, I am grateful to George Henderson, archivist at Queen's University in Kingston, Ontario, which houses the Wynne-Edwards Collection. George's enthusiasm and interest were incredibly important in moving this project forward despite its ever-increasing scope.

Outside the structure of universities and archives, there are family members to thank—both mine and the Wynne-Edwards family. While conducting research at the archives at Queen's University, I was very fortunate to spend time with Dr. Janet Sorbie (V. C. Wynne-Edwards's daughter and a member of the medical school faculty) and Dr. Katherine Wynne-Edwards (Wynne-Edwards's granddaughter and a behavioral endocrinologist on the faculty in the biology department at Queen's). They were generous and welcoming and helped me to get a better sense of Wynne-Edwards's early research in Canada while he was a professor at McGill.

Another family to thank is my professional family at the University of Minnesota. I cannot adequately express how fortunate I am to be at a university where I have constant interaction with historians of science, philosophers of science, and biologists, all engaged in penetrating analysis of the fundamental questions of science. I am particularly grateful to Michel Janssen, whose friendship, sense of humor, and encyclopedic knowledge of Bob Dylan have helped me maintain proper perspective, and to Alan Shapiro, whose guidance and mentorship have allowed me to pursue this research with confidence and freedom. I am also grateful to the Biology

Interest Group at the Minnesota Center for Philosophy of Science, especially Ken Waters, Alan Love, Mike Travisano, and Ben Kerr. The members of this group read drafts of various chapters and provided valuable feedback on the structure and argument presented in chapter 7.

Finally, I thank my own family: my parents, Ronald and Kathryn Borrello, who have always supported me in my varied pursuits. They have encouraged me all the way through, and for that I am eternally grateful. The last acknowledgment is perhaps the most important, and that is to my wife, Regina. She has been there from the beginning of this project and has read more drafts than anyone, heard more versions, and listened to more descriptions—and the work is better for it. I could never have done it without her consistent love and support.

This manuscript incorporates material from my previously published articles: "Synthesis and Selection: Wynne-Edwards' Challenge to David Lack," *Journal of the History of Biology* 36 (2003): 531–566; "Mutual Aid and Animal Dispersion: An Historical Analysis of Alternatives to Darwin," *Perspectives in Biology and Medicine* 47 (2004): 15–31; and "The Rise, Fall and Resurrection of Group Selection," *Endeavour* 29 (2005): 43–47.

Introduction

Since the publication of *On the Origin of Species* in 1859, there have been questions and debates about Darwin's proposed mechanism of evolution—natural selection. Some critics of the theory have argued that Darwin merely applied the prevailing capitalist ethos of Victorian England to the natural world in an attempt to naturalize, and thereby legitimize, the system in which he himself lived and thrived. Others (most famously, perhaps, philosopher Karl Popper) have argued that the mechanism suggested by Darwinism was merely a tautology: survival of the fittest means that those who are fittest are the survivors, hence the empty phrase "the survival of the survivors." This question of tautology has been addressed extensively by biologists and philosophers of biology, and Popper ultimately changed his mind about natural selection. Historians and sociologists of science continue to explore how far prevailing social structures and attitudes influence theory development and acceptance. This book will take up another aspect of natural selection. I am interested in examining the role of group selection in evolutionary theory since Darwin, with a particular emphasis on what happens to this idea in the twentieth century. To accomplish this task in the most illuminating and focused manner, I will use the career of British naturalist Vero Copner Wynne-Edwards as the structural spine of my story. Wynne-Edwards is the perfect vehicle because it was his work that focused the attention of biologists, and especially ecologists, on the ill-formed question of the level at which natural selection was acting.

The theory of group selection is the idea that competition in nature, which is fundamental to Darwin's mechanism of natural selection, also occurs at a level above the individual. Although most of the discussion about natural selection has been at the level of the individual, and more recently at the level of the gene, there have always been researchers committed to the idea that selection must also act on groups of individuals. This idea has been fundamental to explanations of social organization, altruistic behavior, and even the evolution of intellect and morality. In the course of this narrative, I will trace the history of this idea as its status waxes and wanes in the context of the biological sciences. I believe this history of the idea of group selection will contribute in a substantial way to our understanding of the development of evolutionary theory. This account of the history of twentieth-century biology supports a hierarchical understanding of evolutionary theory—that is to say, an understanding that natural selection occurs at multiple levels. Following the history of group selection, from its origins in Darwin's work through the professionalization and specialization of biology in the twentieth century, highlights some important trends in the biological sciences in general.

The history of group selection has been identified by several of the biologists involved in the debates as an important and as yet unattempted historical project. In his 1983 paper "The Group Selection Controversy: History and Current Status," biologist David Sloan Wilson presented his rationale for pursuing a history of the idea.[1] The criteria Wilson laid out served as a starting point for my project. First, group selection is a striking example of how scientific questions arise from attitudes normally considered outside the purview of science and shows that the development of ideas is not necessarily orderly and progressive. Second, even though the modern concept of group selection lies squarely within the older tradition, the differences that do exist must be clarified, and a historical approach is particularly appropriate. Finally, modern group selection theory is consistent with many of the theories (inclusive fitness, game theory, and reciprocity) that had been treated as rival (and mutually exclusive) explanations over the past thirty years. I heartily concur with Wilson's analysis. The history of group selection presented here will examine the work of biologists beginning with Darwin and concluding with the state of higher-level selection theory in the current literature.

The first period under consideration is the Darwinian era. In chapter 1

I use Michael Ruse's 1980 article "Charles Darwin and Group Selection" as a foil for my own interpretation of Darwin's work.[2] Contra Ruse, I will present my argument that in both *On the Origin of Species* and *The Descent of Man* Darwin explicitly allows natural selection to act at levels above the individual.[3] Of course Ruse is not alone in characterizing Darwin as an individual-level selectionist. In his 1998 book *Darwinism's Struggle for Survival*, philosopher Jean Gayon dedicated an entire chapter to what he describes as "this extremely delicate question" and continues, "The virulence of recent controversies over 'group selection' shows that this is a fundamental theoretical question as open today as it was in 1859."[4] Gayon essentially argues that Darwin was indeed an individual-level selectionist, though neither Wallace nor Spencer was.[5]

In chapter 2 I continue the analysis of evolutionary theory by examining some of the major debates over natural selection that occurred in the late nineteenth and early twentieth centuries. These two beginning chapters will set the stage for the early work of Wynne-Edwards on nonbreeding behavior in fulmars and other seabirds that will be presented in chapter 3. In his earliest papers (ca. 1927), Wynne-Edwards showed an interest in the theoretical questions arising from nonbreeding behavior of adult birds that did not fit into the standard Darwinian account of the individual's constantly striving to produce offspring. These papers foreshadowed the debate with David Lack that began in the 1950s.

The developing debate with Lack and the influence of the modern synthesis will be the subject of chapter 4. The approach to higher-level selection takes an interesting turn at this time as a result of the work of the population geneticists. The early mathematical models of Sewall Wright and Theodosius Dobzhansky were of particular significance for Wynne-Edwards. In his seminal *Genetics and the Origin of Species*, Dobzhansky introduced the paradox of viability.[6] In discussing the necessary level of variation that a population must maintain in order to remain viable, Dobzhansky wrote: "Evolutionary plasticity can only be purchased at the ruthlessly dear price of continuously sacrificing individuals to death from unfavorable mutations."[7]

The quotation above is evidence that Dobzhansky was focusing on a group-level or lineage-level trait, evolutionary plasticity or evolvability, and was interested in its importance to evolutionary theory.

With the advent of the modern synthesis, Wynne-Edwards recognized a fundamental shift in the mode of thought about evolution that was more in line with his own interests. In November 1948 he gave a paper to the

Oxford Ornithological Society, "The Nature of Subspecies," in which he discussed the importance of the shift. In his introductory remarks, he cited the work of E. B. Ford on butterflies as well as Dobzhansky's, Ernst Mayr's, and Julian Huxley's core contributions to the development of the modern synthesis. "The fundamental new idea is that populations, rather than independent individuals, are the basic units upon which evolutionary processes act."[8]

Dobzhansky also made interesting claims that the physiology of populations had been entirely neglected and at the same time was perhaps the most essential aspect of the theory of evolution.[9] In his chapter on variation in natural populations, he argued that although the origin of variation was purely physiological, when it is injected into a population it enters into the field of action of factors operating on a different level. According to Dobzhansky, "These factors, natural and artificial selection, the manner of breeding characteristic for the particular organism, its relation to the secular environment and to other organisms existing in the same medium, are ultimately, physiological, physical, and chemical, and yet their interactions obey rules *sui generis*, rules of the physiology of populations, not those of the physiology of individuals."[10]

His emphasis on higher-level selection, which he called group selection, was highly influential for Wynne-Edwards and has previously been underemphasized by authors writing about the synthesis.

The advent of this shift in attention by the population geneticists, followed by the 1954 publication of David Lack's strictly Darwinian (individual selectionist) book *The Natural Regulation of Animal Numbers,* spurred Wynne-Edwards to continue his work formulating a theory of group selection that was consistent with his field experience. The result of this work was a paper presented at the Eleventh International Ornithological Conference that introduced Wynne-Edwards's theory of group selection and rejected the standard Darwinian account represented in the work of David Lack.

The publication of Wynne-Edwards's *Animal Dispersion in relation to Social Behavior* in 1962 is widely recognized as the spark that ignited the contemporary debate over group selection theory, and the reception of this major work will be the subject of chapter 5. In *Animal Dispersion*, Wynne-Edwards expanded his theory based on his work on social behavior in birds and applied it to all animal groups. In his theory of group selection, Wynne-Edwards identified a wide variety of group-level adaptations: reproductive rate, foraging strategy, and strict population localization.

Despite the acknowledgment of the impact of the 1962 book, Wynne-Edwards often remains a footnote in the development of higher-level selection theory. Indeed, in a recent book by biologist Lee Alan Dugatkin, *The Altruism Equation: Seven Scientists Search for the Origins of Goodness*, many of the characters discussed here (including Charles Darwin, Petr Kropotkin, W. C. Allee, and William D. Hamilton) are treated in depth with respect to their contributions, but Wynne-Edwards merits only a single reference.[11]

In chapter 6 I will concentrate on the two most important critics of Wynne-Edwards's theory. First, I will present David Lack's challenge to group selection in *Population Studies of Birds*. This confrontation is particularly interesting because despite the received view that Lack clearly carried the day, closer analysis of the argument reveals that the data supporting Lack's position were surprisingly thin. In combination with the argument presented by George C. Williams in his now classic *Adaptation and Natural Selection*, however, Wynne-Edwards's theory of group selection suffered a mortal blow. This chapter will also engage some of the broader social context that influenced the reception of group selection theory. The 1960s witnessed the rise of the environmental movement and concerns about human population growth. Wynne-Edwards's arguments, presenting group selection as a mechanism of evolutionary restraint on population growth, struck a chord with a number of these groups. Finally, building on the work of historian Richard Burkhardt, I will discuss the ways group selection theory was handled by the ethologists and sociobiologists in their attempts to provide evolutionary accounts of social behavior.

The fate of Wynne-Edwards and his theory of group selection in the wake of the criticisms above and in the context of an increasing emphasis on the gene as the unit of selection is the focus of chapter 7. The increasing involvement of the philosophers of biology in the units of selection debate and the failure of Wynne-Edwards's second book, *Evolution through Group Selection*, round out the chapter and illustrate how the debate moved beyond Wynne-Edwards.

In the concluding chapter I describe Wynne-Edwards's continued attempts to advocate his theory and direct attention to the importance of group selection as an evolutionary mechanism. As we shall see, although biologists remained unconvinced by his formulation of group selection, there was an increasing acceptance of a hierarchical approach to evolutionary theory. In this chapter I will also discuss the role of ideology in the debate over group selection. The varied notions of a "group"—from a tightly organized whole that suppresses the individual to a loose aggregate of individuals who cooperate and sacrifice for mutual benefit—have clearly influenced the arguments on both sides of this contentious issue.

How did Wynne-Edwards develop his theory? What questions was he attempting to answer that others had not resolved? In the course of this history I will trace out the development of the theory in Wynne-Edwards's early work and attempt to identify the social behaviors, the existing theories, and the broader questions about population regulation that led him to his theory of group selection. I will also place Wynne-Edwards and his theory within the context of the developments of evolutionary biology that took place throughout the twentieth century.

CHAPTER ONE

Charles Darwin and Natural Selection

Debates over the nature and power of Darwin's primary mechanism of evolutionary change, natural selection, began with the publication of *On the Origin of Species* and continue into the present. In the sixth chapter of the first edition, Darwin addressed various difficulties his theory faced. These challenges included an oblique reference to the idea that is the subject of this story—the notion that natural selection may act at a level above that of the individual. As the study of biology has developed in the twentieth century, applications of natural selection to levels below the individual, particularly the level of the gene, have become increasingly common and comparatively unproblematic. Applications of Darwin's mechanism in the opposite direction were initially accepted as unexamined claims about the "good of the species," but as these genetic explanations became more prevalent, the higher-level explanations were increasingly seen as suspect.

Sterile Hybrids and Neuter Castes

The idea that natural selection acted at a level above that of the individual was a challenge to Darwin's theory from the beginning. Darwin himself recognized the difficulty of explaining the existence of the neuter castes of social insects by individual selection, as well as the fact of hybrid sterility.

By the end of the century, the neuter insects became the locus of an important debate over the inheritance of acquired characteristics between the neo-Darwinians, represented by August Weismann, and the neo-Lamarckians, led by Herbert Spencer.[1] Darwin's own explanation, as presented below, was based on natural selection acting on traits that were beneficial to the colony. With regard to hybrid sterility, Darwin acknowledged that sterility could be of no possible benefit to the hybrid individual and so must be incidental to other acquired differences. The benefit of the sterility of hybrids accrues not to the individual, according to Darwin, but rather to the species whose integrity is maintained (see fig. 1).

Despite the conceptual ambiguity presented here, historians of science have let this potentially fertile problem lie fallow. The only explicit historical treatment of group selection in Darwin's work was a 1980 article in the *Annals of Science* by philosopher of biology Michael Ruse.[2] His "Charles Darwin and Group Selection" was written in the midst of the sociobiology debate of the mid-1970s and early 1980s. Ruse argued that despite the claims of sociobiology's detractors, Darwin was not sympathetic to the idea of group selection except in a very few particular cases. Ruse achieved this narrow interpretation of Darwin by using modern definitions of individual and group selection that do not apply in the context of the nineteenth century. Individual selection, according to Ruse, is "selection which in some sense affects an individual's reproductive interests. This could be directly through the individual, or indirectly in some way: For instance, by kin selection, where an individual's interests are furthered through close relatives."[3]

The inclusion of kin selection here is dubious. Although Darwin was certainly aware that leaving progeny was important to the struggle for existence, kin selection in the modern sense quantifies relatedness and the benefits of aiding kin in a way that was not possible before the development of classical genetics. Furthermore, it is formally selection between groups of kin.[4]

I believe that Ruse's restrictive reading of Darwin is off the mark. This is evident from various passages from both the *Origin* and the *Descent* that more accurately present Darwin's own position with regard to selection acting at a level above the individual. The following often-quoted passage, from chapter 3 of the *Origin*, titled "The Struggle for Existence," illustrates the breadth of action that Darwin assigned to the mechanism of natural selection: "I should premise that I use the term struggle for existence in a large and metaphorical sense, including *dependence of one*

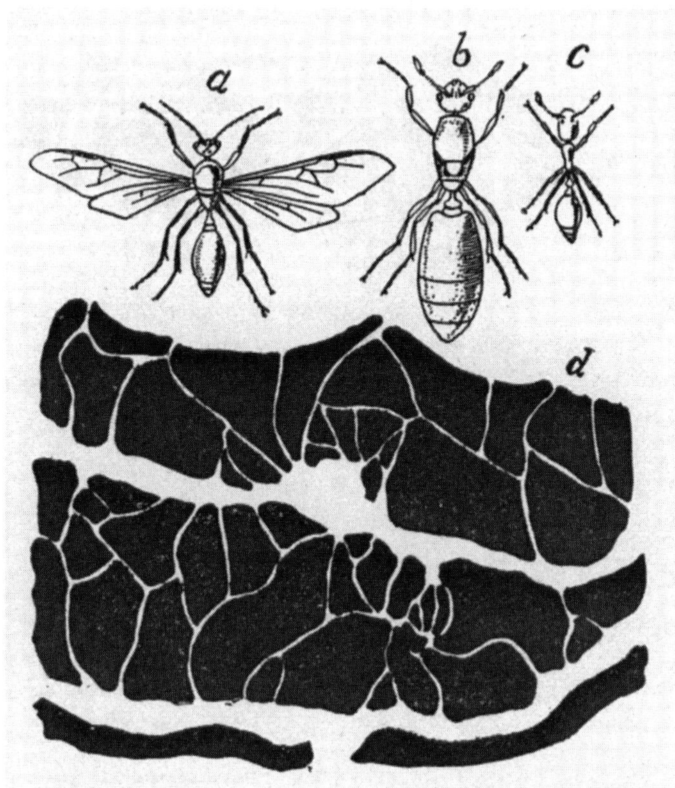

Fig. 245.—The ant, *Solenopsis fugax*: *a*, Male; *b*, dealated female; *c*, worker; *d*, portion of nest showing broad galleries of the host ant intersected by the tenuous galleries of *Solenopsis*, the thief ant. (After Wasmann. See account on page 375.)

FIGURE 1. Social insects. David Starr Jordan and Vernon Kellogg, *Evolution and Animal Life: An Elementary Discussion of Facts, Processes, Laws and Theories relating to the Life and Evolution of Animals* (New York: D. Appleton, 1907).

being on another and including (which is more important) not only the life of the individual, but success in leaving progeny."[5]

Later in the *Origin*, Darwin's statements on higher-level selection dealt mostly with the social insects. Darwin recognized the difficulty that the neuter insects, with their distinct morphology and habits, presented for his theory, and in typical Darwinian style he did his best to explain and defuse this potentially devastating case.

> How the workers have been rendered sterile is a difficulty; but not much greater than that of any other striking modification of structure; for it can be shown that some insects and other articulate animals in a state of nature occasionally become sterile; and if such insects had been social and it had been *profitable to the community* that a number should have been annually born capable of work, but incapable of procreation, I can see no very great difficulty in this being effected by natural selection.[6]

The passage above illustrates Darwin's commitment to the mechanism of selection despite the lack of a clear theory of heredity. The question of the evolution of the neuter insects became fundamental to the ongoing debate over the inheritance of acquired characteristics. Both Lamarckian supporters of use inheritance and neo-Darwinian selectionists used the case to show the insufficiency of the other side's theory. I offer the following lengthy quotation from the *Origin* to demonstrate Darwin's position with regard to the evolution of various castes among the social insects.

> I believe that natural selection, by acting on the fertile parents, could form a species which should regularly produce neuters, either all of large size with one form of jaw, or all of small size having jaws of widely different structure; or lastly, and this is our climax of difficulty, one set of workers of one size and structure, and simultaneously another set of workers of a different size and structure; a graduated series having been first formed, as in the case of the driver ant, and then the extreme forms, from being the most useful to the community, having been produced in greater and greater numbers through the natural selection of the parents which generated them; until none with an intermediate structure were produced.
>
> Thus as I believe, the wonderful fact of two distinctly defined castes of sterile workers existing in the same nest, both widely different from each other and from their parents, has originated. We *can see how useful their production may have been to a social community of insects,* on the same principle that the division of labour is useful to civilised man.[7]

Even though these passages come from the chapter in the *Origin* titled "Instinct," there is no explicit reference to the inheritance of instinct. Rather, Darwin described the morphological traits of the neuter castes and explained them in terms of their usefulness to the community. In an excellent recent review of this problem, historian Thomas Dixon has argued that Darwin's "community selection explanation of the evolution of well adapted neuter insects provided one important example of a case

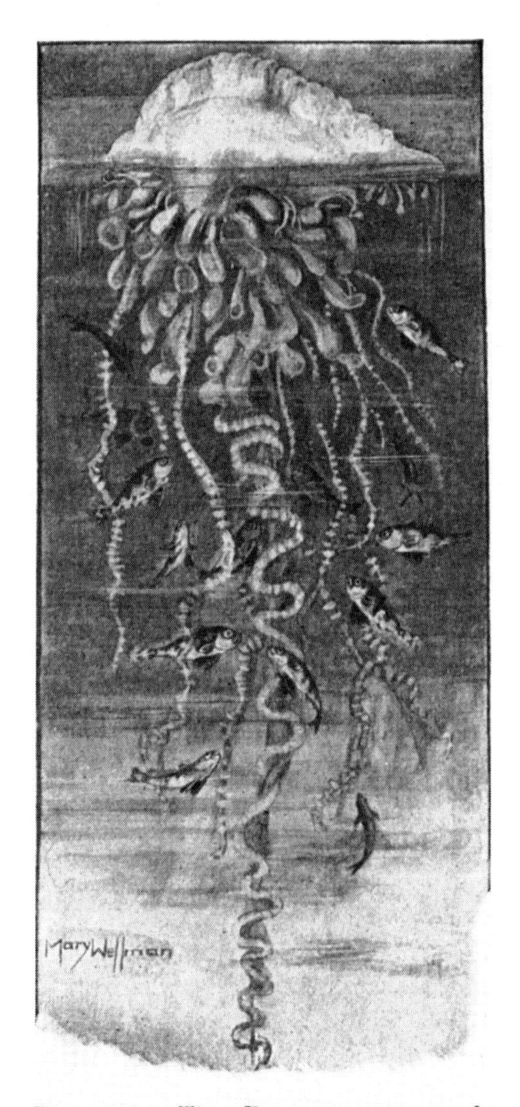

FIG. 228.—The Portuguese man-of-war, *Physalia*, with men-of-war fishes, *Nomeus gronovii*, living in the shelter of the stinging feelers. (Specimens from off Tampa, Fla.)

FIGURE 2. Colonial organism. David Starr Jordan and Vernon Kellogg, *Evolution and Animal Life: An Elementary Discussion of Facts, Processes, Laws and Theories relating to the Life and Evolution of Animals* (New York: D. Appleton, 1907).

that could not be explained by Lamarckian inheritance of modifications produced by use and disuse."[8] Dixon's analysis, consistent with the account I develop here, challenges the characterization of Darwin by scholars such as Helena Cronin, who present "those twentieth-century biologists who invoked group selection as departing from 'the individual-level orthodoxy of Darwin, Wallace and their contemporaries.'"[9] In the case of the social insects, however, instinct was clearly recognized as an important factor in the evolution of the social systems. This idea was more carefully developed in *The Descent of Man*, which I will discuss below, but the following quotation gives some indication of Darwin's position in the *Origin* with regard to instinct in the social insects: "Thus, I believe it has been with social insects: a slight modification of structure, or instinct correlated with the sterile conditions of certain members of the community has been *advantageous to the community*: consequently the fertile males and females of the same community flourished, and transmitted to their fertile offspring a tendency to produce sterile members having the same modification."[10]

It is clear from a careful reading of the selections from Darwin's work that he indeed conceived of the mechanism of natural selection as functioning at the level of the community. This is especially clear in the case of the social insects but was also part of Darwin's theory with regard to communal organisms such as the Portuguese man-of-war and the coral polyp and also true of higher social animals.

The final passage from the *Origin* comes from chapter 6, "Difficulties on Theory." In this chapter Darwin discussed various phenomena that he recognized as potentially contradictory to his theory. Through the broad application of the mechanism of natural selection—that is to say, application to the group rather than the individual—Darwin's theory can encompass even the most self-destructive of "adaptations." "We can perhaps understand how it is that the use of the sting should so often cause the insect's own death: for if on the whole the power of stinging be *useful to the community*, it will fulfill all the requirements of natural selection, though it may cause the death of some few members."[11]

Social Insects and Social Instincts

In *The Descent of Man,* we see Darwin shifting emphasis from the social insects to the social instincts. Generally he continued to use social insects for the model of the evolution of social instincts, but he also included the social behavior of primates and other higher animals. This shift in emphasis acted as an accelerant to the debates generated by Darwinian theory. In *The Descent of Man*, Darwin was explicitly drawing the connection between the moral faculties of man and the social instincts of the lower animals.

Darwin's most straightforward presentation of the evolution of the social instincts came in chapter 3, "Moral Sense." In this passage Darwin argued that the inheritance of the social instincts was of the utmost importance to the later development of the human society and, furthermore, that the development of these instincts was for the good of the community over and above the advantage to the individual. "Finally, the social instincts which no doubt were acquired by man, as by the lower animals, *for the good of the community*, will from the first have given him some wish to aid his fellows, and some feeling of sympathy."[12]

Here we see Darwin's explanation of the human need to offer aid to another in terms of the selective benefit this behavior confers on the community, in the same way that the existence of the sterile caste maintains the selective advantage of the hive. In the next chapter, "On the Manner of Development of Man from Some Lower Form," Darwin pointed out that in the case of the social animals, selection acting at the level of the community could have indirect effects on individuals: "With strictly social animals, natural selection sometimes acts indirectly on the individual, *through the preservation of variations which are beneficial only to the community.* A community including a large number of well-endowed individuals increases in number and is victorious over other and less well-endowed communities; although each separate member may gain no advantage over the other members of the same community."[13]

Darwin went on to illustrate the point with the example of the social insects, describing pollen-collecting behavior and the sting of worker bees in addition to the jaws of the soldier ants. According to Darwin's argument, these apparatuses and behaviors were of no direct advantage to the individual; rather, they served the community and were maintained by natural selection acting on the level of the community.

The final passage from the *Descent* that I will include here illustrates the importance Darwin assigned to the social instincts. "All this implies some degree of sympathy, fidelity and courage. Such social qualities, *the paramount importance of which to the lower animals is disputed by no one,* were no doubt acquired by the progenitors of man in a similar manner, namely, through natural selection, aided by inherited habit."[14]

It follows that if these instincts are as important to the evolution of social groups as Darwin insisted, and if the selection of these instincts often occurs at a level above that of the individual, then higher-level selection is an important factor in Darwinian evolutionary theory. In this way Darwin explicitly equated selection among communities with natural selection—a very different stance from that of later authors, like George C. Williams, who equate only individual selection with natural selection (as I will discuss in chapter 6). Despite Ruse's claim to the contrary, it is apparent that for Darwin there was at least some possibility for his mechanism to function above the level of the individual. At best, Ruse is using a modern definition of kin selection retroactively in evaluating the past. As worst, he is misdefining kin selection in order to "rescue" Darwin from the "error" of group selection.

Darwin's ambiguity on this question of levels at which the mechanism might work contributed to an intellectual environment where claims about "the good of the species" or "the benefit of the community" were accepted unexamined. From the end of the nineteenth century through the first decades of the twentieth, biologists' characterization of this adaptation or that behavior, which had clearly evolved for the benefit of the species or community or group, were generally accepted.

Early Context: The Darwinian Revolution?

Despite a generation of scholarship dedicated to examining the state of Darwinian theory at the end of the nineteenth century, there remained deep disagreements on some very basic points. To what extent was Darwin's work an echo of the Victorian capitalist ethos? What was the actual influence of Malthus on Darwin? What is the meaning of evolution? Was Darwinian theory progressive? These questions, among many others, have fueled the Darwin industry for decades.[15]

Peter Bowler's *The Non-Darwinian Revolution: Reinterpreting a Historical Myth* provides an excellent analysis of the complexity of the situation with regard to evolutionary theory in the period under consideration. Bowler has spent his career examining the varied evolutionary ideas that were being espoused throughout the community of evolutionary thinkers. Bowler's point (and one that is well taken) is that most evolutionists at the end of the nineteenth century and the beginning of the twentieth were not Darwinians, often despite their claims to the contrary.

In *The Non-Darwinian Revolution*, Bowler described the continuing influence of Lamarck's ideas about the importance of use and disuse, Ernst Haeckel's idea of recapitulation—which was closely linked to the idealist and transcendentalist origins of the developmental view of nature—and other scientific and philosophical concepts to illustrate the intellectual confusion that reigned during this period. Bowler also dedicated a chapter to social Darwinism, which he argued was often closer to social Lamarckism. (Bowler argued that if social thinkers wanted people to strive to get ahead, Darwinian theory gave them no grounds to make any effort. Either they had the advantageous traits or they did not. The Lamarckian notion of inheritance of acquired characteristics made quite a bit more sense.) Hence the ultra–social Darwinist Herbert Spencer, as classically presented in Richard Hofstadter's *Social Darwinism in American Thought,* is a misconception. According to Bowler, Spencer's ideas about evolution's improving society were often more Lamarckian than Darwinian.

By 1909, the centennial of Darwin's birth and fifty years after the *Origin*, almost everyone agreed that life evolved, but there was no such agreement on a mechanism. There remained the unsolved problems of variation and heredity, and further complicating matters, physicists' and geologists' estimates of the age of the earth were not providing sufficient time for Darwin's gradualistic account. Even the concept of species was still vague, which created problems for evolutionary theory at the most fundamental level. Meanwhile, new fields in the life sciences arose as older areas of study were modified or abandoned. Weismann and Hugo de Vries, although they themselves did not do research in the field, saw the solution to the problem of inheritance in colloidal chemistry, the study of large molecules that were the basis of life and of how to identify, understand, and analyze them. Physiology, and the work of Jacques Loeb among others, was reshaping biology along mechanistic and experimental lines.[16] Major advances had occurred in plant hybridization techniques. The opening of the American West led to the discovery of major fossil fields and fueled rapid growth of paleontology. The biometricians, led by Darwin's cousin Francis Galton, developed techniques for analyzing inheritance and correlating traits.[17]

The neo-Darwinians were matched by neo-Lamarckians in nearly all of these pursuits. American paleontologists such as Edward D. Cope and Alpheus Hyatt did not see proof of natural selection in the fossil record and maintained a belief in the inheritance of acquired characteristics. Experimental physiologists Paul Kammerer and Charles-Édouard Brown-Séquard performed experiments that supported Lamarckian claims of the inheritance of acquired characteristics. These disputes provide the context for Petr Kropotkin's development of his theory of mutual aid, which I will discuss in chapter 2. In the next section we will examine some general biological texts from the turn of the century in order to gain greater insight into the state of Darwinian theory and, more specifically, the application of natural selection above the level of the individual.

Weismann's Germplasm

August Weismann's collection of lectures, *The Evolution Theory*,[18] provided a transition from the consideration of higher-level selection in general to a more specific treatment of selection in the case of the social insects. This two-volume work consists of thirty-six lectures given at Freiburg im Breisgau over almost as many years. Weismann's lectures presented the neo-Darwinian position disparaged by the followers of Spencer at the turn of the century and described the all-sufficiency of natural selection as the mechanism of evolution.

Besides offering an engaging history of evolution that began with Empedocles and traveled through the works of Goethe and Erasmus Darwin, the first volume introduced Darwinian theory, as well as Weismann's own theory of the germplasm.

Two lectures in the second volume of *The Evolution Theory* are of particular relevance here. Lectures 23 and 24, both of which treat the inheritance of functional modifications, reveal Weismann's position on the role of selection in the evolution of social insects. In the course of his lecture Weismann asked about the behavior and morphology of the neuter insects: "How can all these peculiarities have arisen, since the workers do not reproduce, or do so only exceptionally, and, in any case, are incapable of pairing, and therefore—among bees at least—only produce male offspring?"[19]

Weismann went on to answer his own question, asserting that obviously it could not be through the transmission of the effects of use and disuse, because the workers had no offspring to whom anything could be transmitted.

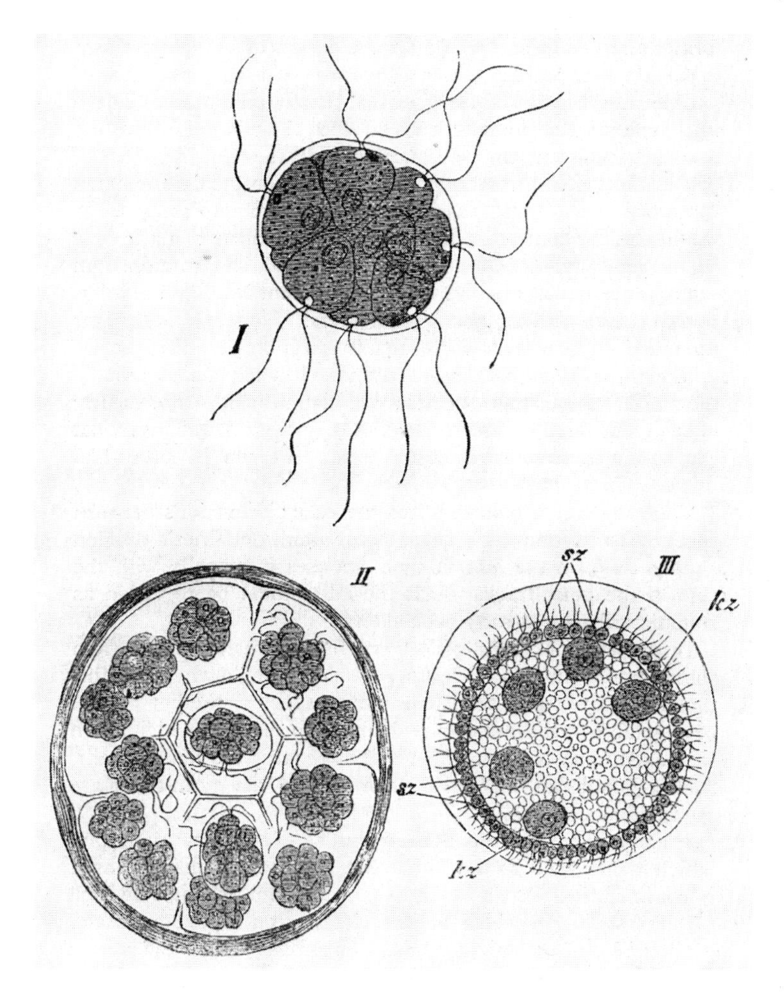

FIGURE 3. Superorganism? August Weismann, *The Germ-Plasm: A Theory of Heredity* (New York: C. Scribner's Sons, 1893), 214.

Weismann continued his explanation by asserting that these varied forms of neuter insects, which exhibited complex behaviors, were the result of natural selection. Essentially, Weismann argued that the inhabitants of the hive were members of the same family and that selection acted on the sexual members of the hive that produced the better workers. In the case of bees, selection acted on the individual queen. Today this would be recognized as family selection if the queen is singly mated, but if she is multiply mated, or joined by other queens, it is a type of kin selection that has been shown to be among-group selection.[20]

Given Weismann's theory of the continuity of the germplasm, it is perhaps not surprising that he would offer this explanation. Nevertheless, one might argue that there is room for some selective force to act on the level of the group, given the following passage from the same chapter. "A colony whose queen was unsatisfactory in this respect [producing good workers] could not hold its own in the struggle for existence, and only the best colonies and the best hives would survive, that is, through their descendants."[21]

It appears from the passage above that, although Weismann is committed to the position that selection acts on variations in the germplasm, somehow the fitness of the hive or colony is an important factor along the lines of Thomson's sieves, discussed below. The idea of higher-level selection is often presented in the context of more general studies of Darwinian theory of the period. The social insects, however, were often invoked as the model for studies of complex behavior and sophisticated social systems that might require higher-level selection. With this in mind, chapter 2 will focus on the varied positions taken by evolutionary biologists considering the social insects, and on the concepts of the superorganism and of mutual aid. Before turning to a discussion of these concepts, I will analyze some general treatments of the state of evolutionary theory. I have selected these works as representative of the very broad interpretation of evolutionary theory common to the period and specifically for their treatment of higher-level selection.

Higher-Level Selection in General Biological Texts

The first of these texts, *Evolution and Animal Life*, by David Starr Jordan and Vernon Lyman Kellogg, was published in 1907.[22] David Starr Jordan is well known (especially at Indiana University, where the biology building bears his name) as an ichthyologist and field naturalist of the first order. His work led to the naming of more than 2,500 species of fish. After serving as president at Indiana University for five years, Jordan became the first president of Leland Stanford Junior University. Vernon Kellogg joined the faculty at Stanford in 1894 as a professor of entomology and became head of the department the next year. This text, according to the authors, attempted to "provide a lucid account of the processes of evolution so far as they are understood." The book did not discuss the evolution of plants because both of the authors were trained as zoologists. Nevertheless, *Evolution and Animal Life* is useful as an example of a general biology text of the period. The material covered was generated from the lectures Jordan and Kellogg used in introductory biology courses at Stanford University.

The subtitle of this work, *An Elementary Discussion of Facts, Processes, Laws and Theories relating to the Life and Evolution of Animals,* gives one a fair sense of its scope. Jordan and Kellogg began with a chapter on the definition of evolution and concluded twenty-one chapters later with a discussion of man's place in nature. Although the authors were certainly supporters of Darwin, their position with regard to the mechanism of natural selection was not absolute.

Chapter 4, "Factors and Mechanism of Evolution," included the subsection "Lamarckism and the Inheritance of Acquired Characters." In this chapter, Jordan and Kellogg asserted that, although they considered natural selection to be the predominant mechanism in the course of evolution, Lamarckism must not be omitted. This was especially true in the context of development. The authors also included an entire chapter (11) on the inheritance of acquired characteristics. The message with regard to Lamarckism here is less clear. Jordan and Kellogg seem to want to downplay the importance of Lamarckian ideas as much as possible; however, they also want to avoid any direct affiliation with the *Allmacht* stance of the neo-Darwinian Weismann. Jordan and Kellogg went on to discuss the role of mutual aid and communal life in chapter 18, which provides some indirect insight into their position with regard to selection acting on groups.

In this chapter the authors treat the social insects but do not make specific reference to the role of natural selection in the development of insect social systems. They do, however, emphasize the importance of the group, especially in the case of bees. For example, the discussion of the worker bee concentrates primarily on its morphological differences from the other members of the hive; however, it also describes the contribution of the worker as more important to the community than to the individual itself. "And all work done by the workers is strictly work for the whole community; in no case does the worker bee work for itself alone; it works for itself only in so far as it is a member of the community."[23]

At the end of the chapter, in which they include examples of a number of dependent relationships, including symbiosis and parasitism, Jordan and Kellogg acknowledge the importance of communal and social life in the animal world and, more specifically, the important advantages it provides in the struggle for existence. "The advantages of communal or social life, of cooperation and mutual aid, are real. The animals that have adopted such a life are among the most successful of all the animals in the struggle for existence."[24]

Although there is never any discussion of the mechanism at work here, it appears that the authors are quite close to Darwin's own position with regard to selection working at higher levels. Their social context may have played a role in their closing comments for this chapter. After extolling the virtues of community and cooperation, the American zoologists, teaching at Stanford University in the great western region of the United States, felt compelled to qualify their claims. "Its great advantages are, however, balanced by the fact that mutual help brings mutual dependence. The community or society can accomplish greater things than the solitary individuals, but cooperation limits freedom, and often sacrifices the individual to the whole."[25]

Jordan and Kellogg, though convinced of the importance of the group in nature, would not deny the equally important role of the individual, especially in a free and democratic society like the United States.

The following year Vernon Kellogg published *Darwinism Today*.[26] In this work he attempted to provide a balanced discussion of the varied evolutionary theories of the time and included some interesting passages on the efficacy of the mechanism of natural selection. Here again he attempted to locate himself close to Darwin with respect to the role of natural selection in evolution. Even though Kellogg admired the work of Weismann and other neo-Darwinians, he cautioned that they had gone too far.

Kellogg argued that although Weismann had provided substantial refutation or alternative explanations for every example of inheritance of acquired characteristics, he nevertheless had to modify his own position on the *Allmacht* of natural selection in light of the work of Karl Wilhelm von Nägeli and Yves Delage on the "swamping" of favorable variations.[27]

With regard to selection's acting at a level higher than the individual, Kellogg deferred to Thomas Hunt Morgan's argument against natural selection's giving rise to sterile castes in the social insects. In *Evolution and Adaptation* Morgan had argued that the neuter castes were perhaps not adaptive at all and were merely the result of a nonlethal mutation.[28] Initially, the mutation theory provided an account of the origin of new variations that were distinct enough to be of evolutionary significance and were also heritable. Morgan espoused this position, in direct opposition to Weismann and the neo-Darwinian overemphasis on natural selection as the creative force in evolution. Morgan's position on natural selection continued to develop in the first decades of the twentieth century, and by the time he wrote *The Scientific Basis for Evolution* in 1932, he had come to accept natural selection in concert with a modified mutation theory.[29]

Ultimately it is difficult to discern exactly where Kellogg stands on Darwinism. Reflecting the contemporary atmosphere while at the same time constructing it, he considered a multitude of possibilities with regard to evolutionary theory. His book provides an archetypal illustration of the state of affairs described by Bowler in *The Non-Darwinian Revolution*.

The structure of *Darwinism Today* is particularly illustrative of the state of Darwin's theory almost fifty years after the *Origin*. The book begins, inauspiciously for Darwin's theory, with the chapter "Introductory: The Deathbed of Darwinism." The second chapter is dedicated to an exposition of the distinction between evolution and Darwinism, and the following three chapters are headed "Darwinism Attacked." In a section in chapter 4 subtitled "How Real Is Personal Selection?" Kellogg asserts: "I may say baldly that no such vigour of individual selection based on variation in colour, in pattern, in venation and other wing characters, in hairs and in numerous other structural characters, as demanded by the needs of selection theory, is to be detected. . . . in other words, just as much variation exists after enduring the selective rigour of the struggle as existed before."[30]

Here Kellogg's sympathy for Darwin's theory seems to have disappeared. The following chapters (6 and 7) are headed "Darwinism Defended," and then Kellogg presents a discussion (comprising two full chapters) of alternative evolutionary theories. In the concluding chapter, "Darwinism's Present Standing," Kellogg essentially returns to the Darwinian fold and to the position he took with Jordan in *Evolution and Animal Life*. He asserts that despite the ongoing debate over many of the particulars, natural selection must still be considered the final arbiter in descent.

Thomson's Sieve

In James Arthur Thomson's *Concerning Evolution* we get perhaps the most Darwinian picture of evolution considered to this point.[31] Thomson, a professor of natural history at Aberdeen University and the translator of Weismann's work, derived this book from a series of lectures presented at Yale University in 1924. His object, according to the preface, was to show that the evolutionary view of nature and of man provided an enriching and encouraging account, contrary to popular understanding. This was not to be an exclusively Darwinian reading of evolution; in a section headed "Self-Regarding and Other-Regarding," Thomson quoted Spencer on the importance of mutual aid. "As Herbert Spencer said: 'From the dawn of life altruism has been no less essential than egoism. Self-sacrifice is no less primordial than self-preservation.'"[32]

Throughout his book, Thomson emphasized the importance of what he called Darwin's subtlety with regard to the idea of the struggle for existence. He introduced the notion of selective sieves acting on different aspects of an organism and at different levels (sieve of the quest for food, sieve of the physical environment, sieve of the animate environment, sieve of courtship). Although these ideas are not developed into a systematic theoretical structure, they nevertheless indicate Thomson's sympathy for the idea of selection acting at multiple of levels. He made specific reference to selection acting at the level of society in his chapter on organic evolution: "Moreover, under the shelter of society there is a possibility of new departures which would be speedily eliminated by the sieves which apply to ordinary, more or less individualistic, life. At different levels of animal society there will be a different pattern of sieve."[33]

Clearly, altruistic behavior, which would be difficult to explain by a selective sieve operating at the individual level, could be effectively explained given another selective sieve operating at the level of the societal group. The altruistic group, having the higher fitness owing to cooperative effort, would outlive the group of selfish individuals.[34]

Thomson's work was essentially Darwinian—that is to say, selectionist but more teleological than Darwin himself might have liked. Most relevant for our purposes, Thomson clearly supports the action of selection at the group level.

Through the analysis of these selected texts it becomes apparent that Weismann had a far-reaching effect on the development of evolutionary theory in general around the turn of the century. Many biologists, although convinced of evolution and sympathetic to Darwinism, remained skeptical regarding the mechanism of natural selection. How and where the mechanism worked was unclear, and the idea was inconsistently applied. Clearly there was difficulty making progress from Darwin's nineteenth-century heredity theory to the more transitional one represented by Weismann. This difficulty was initially compounded by views like Morgan's and by the challenge that faced theorists attempting to sort out the evolution of the sterile castes of the social insects. In the next chapter we will see how these ideas regarding the social insects and superorganisms laid the foundation for further investigation into the possibility of higher-level selection.

CHAPTER TWO

Social Insects, Superorganisms, and Mutual Aid

Social Insects and the Inheritance of Acquired Characteristics

According to my review of the *Zoological Records*, in the earlier years of the period of interest here (after *The Descent of Man*), work on the social insects was particularly meager. It was not until the 1890s that relevant publications began to appear. Unfortunately, this is also the period that attracted Bowler's attention for the plethora of non-Darwinian ideas floating around, and it was also the period of the Spencer-Weismann debate.

The idea that stirred the most significant debate with regard to the social insects was again the inheritance of acquired characteristics. In the context of this debate some interesting arguments were made that bear directly on the issue of higher-level selection.

At a meeting of the Academy of Natural Sciences of Philadelphia on May 23, 1894, neo-Lamarckian paleontologist Edward D. Cope presented a response to William P. Ball's rejection of the theory of use inheritance as a factor in the social colonies of *Hymenoptera,* presented in Ball's book *Are the Effects of Use and Disuse Inherited?* Cope suggested that the proper explanation of the variety of the neuter members of the social insects "will be probably some day found by paleontologic discovery."[1] He predicted that evidence would be produced of formerly heterogeneous colonies that had slowly evolved to their present differentiated state, "due to the usual process of specialization through use-inheritance."[2]

Cope concluded his argument against Ball with the following assertion: "It is enough for the present purpose to have shown that it is more probable that the basis of the entire community, the original fertile soldier, acquired his characters in the usual way: by use, and that all other forms have been derived from him by inheritance, modified by disuse or degeneracy, under the influence of variations in the food supply."[3]

This short piece by Cope was prompted by an exchange between Ball and Joseph T. Cunningham during February and April 1894, in the journal *Natural Science.*

The episode was concurrent with the Spencer-Weismann debate and was clearly an offshoot of it. It began with a short article by Ball, "Neuter Insects and Lamarckism." In this article, Ball essentially recapitulated Weismann's arguments for natural selection against Spencer. He included some examples from the natural history of the Hymenoptera that could not be explained by use inheritance. Consistent with Weismann's explanation presented above, Ball argued that natural selection had acted on the fertile members of the community and thus brought about the production and preservation of the most useful forms and habits in workers. "The whole circumstances tend to show that Natural Selection has simultaneously effected increasing specialisation in neuters and queens alike in the direction of the division of labour and the more perfect adaptation of each caste to its own particular functions in the community."[4]

In his response to Ball's article, Joseph Thomas Cunningham, lecturer in zoology at East London College and a Spencerian acolyte, claimed as his target the new examples presented by Ball and promised to dismiss them. Nevertheless, the entire first half of his article was dedicated to refuting the arguments Weismann made in his September 1893 *Contemporary Review* paper, "The All-Sufficiency of Natural Selection," which initiated the Spencer-Weismann debate.

Ultimately, Cunningham addressed Ball's new examples and delivered the most rhetorically convincing of arguments for rejecting the selectionist account of the evolution of the social insects. Cunningham crowed near the end of his piece, "So far, then, from being a triumphant proof of the all sufficiency of natural selection, the peculiarities of social polymorphic insects, rightly regarded, offer the strongest support of Lamarckian principles."[5]

The relevance of this exchange between the neo-Darwinian Ball and the neo-Lamarckian Cunningham is the same as that of the Spencer-Weismann debate. For the neo-Lamarckian, selection had not been an important factor at any level. Cunningham went so far as to say, "I believe that some day it will be generally admitted that the whole natural selection doctrine is, in the etymological sense, preposterous."[6] On the other hand, Ball's position with regard to selection's acting at a level above the individual was a clear reflection of Weismann. He appeared to allow for some kind of fitness evaluation at a higher level, although traits are selected in the germplasm of the queen. He wrote, "Originating, as we may suppose, in exceedingly remote ages, and forming an advantageous basis or accompaniment of social forms of evolution, the peculiarity or susceptibility has been evolved and utilised to an extent which has allowed special scope for many important differences and complexities which otherwise could not have arisen."[7]

Interesting terms in this passage include social forms of evolution and special scope. The precise meaning of these phrases is unclear, but one could argue that they acknowledge some role for natural selection at a level above the individual, particularly given the historical conceptual context.

The Physical and the Psychical

The rest of the papers I reviewed for this study moved away from the use transmission issue and introduced some of the other factors that were a part of the discussion of the evolution of the social insects.

In the *Report of the British Association for the Advancement of Science,* Charles V. Riley, a biologist at the University of Washington, presented a more difficult case. In "On Social Insects and Evolution" Riley was, perhaps not surprisingly, both Darwinian and non-Darwinian. He argued at one point that "the variations in social insects have been guided by natural selection among colonies."[8] Here he rightfully claimed consistency with Darwin and allowed that selection was working at the colony level. Later in the paper, however, in a response to what he regarded as an overemphasis on cell theory, he wrote: "But it is difficult to believe, in light of the facts which are known concerning social insects, that the different kinds of ids and determinants which are thus conceived to characterise the germ have not been impressed upon it as a consequence of the characters, both acquired and congenital, of the parents."[9]

According to Riley, the use inheritance debate ultimately missed the point. His greater interest lay with the study of psychical evolution. Riley argued:

> The trouble with all theories of reproduction and heredity based solely on embryologic and microscopic methods is, that the essence, the life principle, the potential factors, must always escape such methods. Any theory that will hold must cover the psychical as well as the physical facts—the total of well established experience—and this truth was recognized by Darwin in framing his tentative theory of pangenesis.[10]

Although this emphasis on psychical evolution may sound a bit mystical, Riley made it a fundamental part of his work. In another paper, "Social Insects from Psychical and Evolutional Points of View," presented as a presidential address to the Biological Society of Washington in January 1894, Riley laid out in detail the importance of psychical evolution to the social insects. He discussed the high degree of development of the five senses in insects and offered evidence of insect intelligence as well as the possibility of telepathy as important factors in their psychical evolution.

It is apparent that Riley's picture of evolution, both psychical and physical, included both Lamarckian and Darwinian components. He essentially agreed with Cunningham and Spencer about the role use inheritance and nutrition played in the development of the neuter castes of social insects. But Riley also accepted (if a bit grudgingly) Darwin's notion of selection acting on colonies: "Natural Selection, if it has played any part at all [in the development of the neuter castes], must have done so chiefly in the manner ingeniously suggested by Darwin himself, namely, not as between individuals, but between colonies. The tendency to produce arrested females or sisters doubtless became fixed in some ancestral form through social selection, and is kept up by this and colony selection."[11] Within the notion of social selection, Riley included the development of intelligence, which had been enhanced by use inheritance and education in the hive. These factors, which are not a part of the previous higher-level selection examples discussed here, serve to confound the issue. Riley's distinction between social selection and colony selection is difficult to parse. Furthermore, Riley argued that the time was right for acknowledging animal intelligence and that resistance to the idea was evidence of intellectual sluggishness and human pride. He asserted:

> With all [social insects'] other traits so comparable to those characteristic of human society, they will hardly be charged with the possession or practice of any theology; yet we may look in vain, among all the nations of the earth, unless,

> indeed, among the similarly unblessed aborigines of Borneo and some other lands, for greater self-sacrifice or courage in defending the common weal; for greater loyalty to the sovereign head of the community, not made by divine right, but practically chosen by the commoners; for greater attention or care in the education of the helpless young, or for more harmonious or friendly action between the individuals that form the community. Without form or ceremony they have developed an altruism which with us is believed to exemplify the highest phase of civilization.[12]

This passage from Riley's presidential address provides a clear illustration of the confounding effect that notions of intelligence and morality had on discussions of social insects. In the late nineteenth century the idea of group selection not only is closely linked to the debate over use inheritance but is also integral to Riley's account of psychical evolution and larger discussions of animal intelligence and the existence of morality in nature.[13]

Superorganism: The Group as Individual

The final issue to be discussed in this section on the social insects is the notion of the individual. The concept of individuality continues to occupy contemporary biologists, and with the advent of cloning technology it will only become more vexed. The importance of the study of individuality to the idea of higher-level selection is nicely illustrated by the work of the Harvard zoologist William Morton Wheeler.

In a lecture delivered at the Marine Biological Laboratory at Woods Hole in 1910, Wheeler made a compelling argument for considering the ant colony as an organism. It is important to point out that Wheeler was not merely analogizing. He was arguing along Weismannian lines, presenting the queen as the germplasm and the workers as the soma. He went on to stipulate, in the course of his address, that the division of labor between the two classes of the nutritive worker division and the protective soldier division clearly resembled the differentiation of the personal soma into endodermal and ectodermal tissues.

Wheeler concluded the paper by asserting that we must pay closer attention to the innumerable cases of symbiosis, parasitism, and coenobiosis: "Since in all of these phenomena our attention is arrested not so much by the struggle for existence, which used to be painted in such lurid colors, as by the ability of the organism to temporize and compromise with other organisms, to inhibit certain activities of the aequipotential unit in the interests of the unit itself and of other organisms; in a word, to secure survival through a kind of egoistic altruism."[14]

This interest in the colonial organism was not Wheeler's alone and, I argue, acts as a true confounding factor in the study of ideas of higher-level selection in this period. Wheeler mentioned a two-volume work by Hans Driesch, *Philosophy of the Organism*, whose vitalism he found particularly unscientific.[15] Julian Huxley had written a short work on the subject, *The Individual in the Animal Kingdom*. With regard to the general texts discussed earlier, Jordan and Kellogg's chapter "Mutual Aid and Communal Life among Animals" included discussions of colonial organisms such as the Portuguese man-of-war and asexually reproducing animals such as the hydra. In the *Principles of Biology*, Spencer admitted that individuality is a problem for the biologist; however, he advised, contra Wheeler, that we must "accord the title of individual to each separate aphis, each polype of a polypedom, each bud or shoot of a flowering plant, whether it detaches itself as a bulbi or remains attached as a branch."[16]

As should be clear from the panoply of opinion presented here, there is no way to break the issue down into a clear two-part comparison of those who supported higher-level selection and those who rejected it. I think, however, that examining the various positions represented here will lead to a better contextual understanding of the attitudes toward group selection than the oversimplified picture presented by Ruse, as described in chapter 1.

The predominance of the use inheritance debate, the inclusion of moral and intellectual faculties in the analysis of social insects, and the variety of opinion with regard to the definition of the individual all influence each person's point of view with regard to group selection. Another factor, which has certainly played a role in the contemporary debate over group selection but that I only mention in passing in the body of the chapter, is intellectual or ideological milieu. Daniel Todes's study of Russian naturalists, *Darwin without Malthus,* presents an excellent example here.[17] Furthermore, one of the key individuals in Todes's work, Petr Kropotkin, provides an interesting bridge from the turn of the century to the later work of V. C. Wynne-Edwards.

Petr Kropotkin: A Case Study

Despite nodding familiarity with his *Mutual Aid: A Factor in Evolution*, Petr Kropotkin is perhaps better known to Western scholars in political science departments, as the leader of the international anarchist movement, than in the history of science. This is evinced by a recent article by Joel Schwartz in the *Journal of the History of Biology,* which characterizes Kropotkin as purely an amateur naturalist.[18] On the contrary, Petr Alekseevich Kropotkin cultivated an interest in natural history from an early age and pursued that interest as a member of various expeditions and a contributor to professional societies. On graduating first in his class in the Corps of Pages, he became the czar's *page de chambre,* but he quickly tired of court life. Against his father's advice, he took an opportunity to travel through Siberia as naturalist on a series of commercial and military expeditions. On July 24, 1862, Kropotkin left for Siberia; over the next five years he would travel 50,000 miles and gain experience in geography, geology, and zoology. By the time of his Siberian adventures he had eagerly read, and discussed with his older brother Alexander, the works of various explorers and naturalists including Alexander von Humboldt. During these travels he read Darwin's recently arrived but not yet translated *On the Origin of Species*.

Kropotkin's Siberian travels, supported by the Imperial Russian Geographical Society, led to a year devoted to geography and to the study of the Ice Age formations of eastern Siberia. In 1866 he participated in another of the society's expeditions, which sought an overland route for transporting cattle between Chita and the Lena gold mines. Kropotkin wrote several scientific articles during these expeditions and coauthored the Olekmin-Vitim expedition's scientific report, published in 1873.

When he returned to St. Petersburg from Siberia, Kropotkin took a job with the Ministry of the Interior but continued his scientific career, and he gained membership in the Imperial Geographical Society as a result of his contributions. In 1868 he was awarded the gold medal for his work with the Olekmin-Vitim expedition, and in 1870 he became secretary of the society's section of mathematical and physical sciences.

During this time Kropotkin was also pursuing his association with radical politics. In 1872, hours after delivering a paper on the origins of the Ice Age, he was arrested for trying to popularize revolutionary socialism among workers. While in prison Kropotkin continued his scientific work, and he completed his *Investigation of the Ice Age* there in 1876. That year

FIGURE 4. Petr Kropotkin. Photo by Félix Nadar.

he also escaped from prison and fled to England, where he continued to pursue both politics and science. He published several articles in *Nature* and contributed his knowledge of Siberia to the French geographer Jean-Jacques-Élisée Reclus's *Nouvelle Géographie universelle: La terre et les hommes*. Meanwhile, the "anarchist prince" had also become a leader in the international anarchist movement. He was imprisoned again in France in 1882, sentenced to five years, but was released in 1886 and returned to England.

Kropotkin continued to write about history, politics, and the natural sciences and became a regular contributor to the popular journal the *Nineteenth Century*. This periodical was the venue for his development of his mutual aid interpretation of Darwinian theory. He became a member of the British Association for the Advancement of Science in 1893, investigated remnants of the Ice Age on an expedition to Canada in 1897, and reported to the Royal Geographical Society in London on "the desiccation of Eur-Asia" in 1907. Cambridge University offered him a chair in geology in 1896, but Kropotkin declined because the appointment was contingent on a pledge to abstain from political activity.[19] Kropotkin returned to Russia in 1917 after the czar was deposed, and he worked on his final book, *Ethics,* until his death in 1921.

Kropotkin in Siberia

Although Kropotkin's experience in Siberia can be likened to Darwin's voyage on the *Beagle* inasmuch as both are representative of what Susan Cannon described as "Humboldtian science"—gentleman naturalists collecting observations and information on the fauna and flora of generally unknown regions—the ecosystems they encountered could not have been more different. Darwin's voyage passed through nearly every ecological zone, including some that even resembled the Russian steppes; but he was most influenced by his experiences in the tropics. "The glories of the vegetation of the tropics rise before my mind at the present time more vividly than anything else," Darwin wrote in his autobiography.[20] Kropotkin's travels ranged over the wilds of Siberia, where organisms coped with a wide range of rapidly changing conditions: extreme summer heat and freak blizzards, drought and sudden torrential rainstorms.

Kropotkin's experience was clearly colored by his reading of Darwin. In the first chapter of *Mutual Aid,* he wrote:

> I recollect myself the impression produced upon me by the animal world of Siberia when I explored the Vitim regions in the company of so accomplished a zoologist as my friend Poliakov was. We were both under the fresh impression of the *Origin of Species*, but we vainly looked for the keen competition between animals of the same species which the reading of Darwin's work had prepared us to expect. . . . We saw plenty of adaptation for struggling, very often in common, against the adverse circumstances of the environment, or against various enemies, and Poliakov wrote many a good page upon the mutual dependency of carnivores, ruminants, and rodents in their geographical distribution; we witnessed numbers of facts of mutual support. . . . even in the Amur and Usuri regions, where animal life swarms in abundance, facts of real competition and struggle between higher animals in the same species came very seldom under my notice, though I eagerly searched for them.[21]

In the development of Kropotkin's theory, he wrote that his experience in Siberia influenced his reaction and understanding of Darwin's theory in four significant ways. First, the harsh environmental conditions emphasized the organisms' struggle against abiotic forces. Second, he saw sparse population as the common condition in nature, contradicting the Malthusian aspects of Darwin's theory. Third, where there were large numbers of animals, he was struck by the herds and colonies as cooperative communities struggling against other species and physical hardship. Finally, he observed that under the harshest conditions, where the Darwinian would expect competition to be most severe, the entire group's survival was at risk, and its overall fitness diminished.

The Siberian experience played a role in the development of Kropotkin's ideas about evolution and ecology, just as Darwin's *Beagle* voyage had done in his. Just as Darwin came to doubt the fixity of species, so did Kropotkin, already an evolutionist, become skeptical of the importance of intraspecific competition.

The Theory of Mutual Aid

In *Darwin without Malthus*, Daniel Todes develops the context of the community of Russian naturalists in which Kropotkin further developed his ideas about mutual aid on his return from Siberia. The specific event that led Kropotkin to publish his theory was reading Thomas Huxley's 1888 essay "The Struggle for Existence in Human Society," which represented Huxley at his most Huxleyesque. Ranging through natural history, anthropology, political theory, philosophy, and economics, in this very engaging work he described the role competition has played in human society from prehistoric times up to contemporary Victorian England.

In this essay, which Kropotkin later described as "atrocious," Huxley depicts nature as "on about the same level as a gladiators show. . . . [in this case, however,] the spectator has no need to turn his thumbs down as no quarter is given."[22] In the course of the essay, which ends with a plea for educational reform, Huxley clearly conveys his belief in nature as a zero-sum game. The one natural urge that human society cannot override is the innate urge toward reproduction: "Let us be under no illusion then. So long as unlimited multiplication goes on, no social organization which has ever been devised, or is likely to be devised, no fiddle-faddling with the distribution of wealth, will deliver the society from the tendency to be destroyed by the reproduction within itself, in its intensest form, of that struggle for existence, the limitation of which is the object of society."[23]

Ultimately Huxley's strategy for a successful society was twofold. The first requirement was good products at low prices, and the second was social stability. In the rest of the essay he describes the educational reforms, especially in the sciences and the arts, that would achieve these ends. Huxley's suggestions for educational reform and workers' rights appear quite reasonable. It was essentially the mischaracterization of the state of nature and the argument that flowed from it that Kropotkin so vehemently rejected.

Kropotkin quickly responded to Huxley's article, and although Huxley never fired a countervolley, Kropotkin would spend the next twenty years publishing the developments of his theory in the *Nineteenth Century,* and then in the *Nineteenth Century and After*. This series of articles, the first eight dealing primarily with mutual aid in human evolution and subsequently published as *Mutual Aid: A Factor of Evolution*, followed by seven more articles published between 1905 and 1919, represented Kropotkin's struggle to broaden the scope of Darwinian theory in a manner consistent with Darwin.

Kropotkin and Selection

In "The Theory of Evolution and Mutual Aid," Kropotkin presented the history of the development of Darwin's theory through an analysis of Darwin's correspondence and the changes in the various editions of the *Origin.* He then presented his own theory, as the logical conclusion of Darwin's theoretical development. He argued that Darwin was initially so conservative, with respect to mechanisms other than natural selection, because of a simple paternal predilection for his idea, compounded by a real need to differentiate his theory of evolution from those of Lamarck and Robert Chambers.

In the course of this article Kropotkin addressed the apparent tension between his theory and Darwin's and explained that it was a function of the Western Darwinists' commitment to Malthusian doctrine. The shift of emphasis from Malthusian, individual competition to organisms engaged in a constant struggle for existence, primarily against environmental forces but also against other species, allowed an increased role for mutual aid. Citing Darwin's letters, Kropotkin showed that Darwin also responded to criticism that he had underestimated the importance of the direct action of the environment. Kropotkin wrote, "He gradually came, in an indirect way, to attribute less and less value to the individual struggle inside the species, and to recognise more significance for the associated struggle against the environment."[24]

Kropotkin argued that in the course of the struggle against the environment, species were more apt to practice mutual aid, *and* that cooperative species would increase in numbers and outlast their individualistic rivals. In this scenario, natural selection ceases to be "a selection of haphazard variations, but becomes a physiological selection of those individuals, societies and groups which are best capable of meeting the new requirements by new adaptations of their tissues, organs and habits. It operates largely as a selection of groups of individuals, modified all at once, more or less, in a given direction."[25]

Kropotkin's argument regarding the direct action of the environment led to another important differentiation of his theory from early Darwin and the neo-Darwinians. This was the perceived difficulty for natural selection to establish new species working from random variation. Kropotkin essentially argued that we now know variation is not purely random. The work of the biometricians demonstrated that "whether we take the sizes of the leaves of the same tree, or the stature of several thousand Englishmen at Cambridge University . . . or the contents of sugar in the beetroot, everywhere we find that the laws of variation in organic beings are the same as those with which we are familiar in physical sciences under the name of laws of errors in the theory of probabilities."[26] Kropotkin deduced from this that if indeed we see some deviation from the curve in a particular direction, it must be the result of some permanently acting cause, that is, the direct action of the environment. Furthermore, once there is such a cause, there in no need for an acute struggle between the individuals of the species to preserve the effects of variation. The acting cause itself will accumulate them and increase them in subsequent generations. Finally, Kropotkin pointed to the role of isolation in speciation as a significant development that had led to his theoretical extrapolation from the early Darwin. After quoting passages on the importance of isolation from Darwin's correspondence, Kropotkin asserts:

> Once we admit the successive migrations, in the course of ages, of certain species over several continents (and it seems necessary to admit them, for instance, for the series of ancestors of the wild horse), and once we realise the amount of segregation that ensued, we fully understand the necessary "absence of intermediate forms." And yet it was this absence which so much puzzled Darwin and for which he admitted "extermination" during a severe struggle for life.

With isolation, such an extermination is not necessary; and probably it did not take place at all.[27]

In the second paragraph of the final article, published in 1919, Kropotkin summarized the focus of his series of papers:

A mass of researches having been made on the great fundamental question as to the part played by the direct action of the environment, I analysed them in a series of articles published in this review during the last seven years. Beginning with the evolution of the conceptions of Darwin himself, and most evolutionists about Natural Selection, I next gave an idea of the observations and experiments by which the modifying powers of a changing physical environment were established beyond doubt. Then I discussed the attempt made by Weismann to prove that these changes could not be inherited and the failure of this attempt. And finally, I examined the experiments that had been made to ascertain how far the changes produced by a modified environment are inherited.[28]

What was left of the project was considering what conclusions to draw from this research. Kropotkin argued that Darwinian theory had been pushed to false extremes by his followers. Again he cited the correspondence, as well as Darwin's 1862 work *The Variation of Animals and Plants under Domestication,* in support of his interpretation of the true nature of Darwinian theory. The primary conclusion was that the neo-Darwinians were blinded by Malthusian orthodoxy. This led to his second conclusion, that they were committed to a false polarization into Darwinist and Lamarckian camps. Finally, according to Kropotkin, and despite his own use of experimental evidence in support of his point of view, biologists had lost touch with the naturalist tradition and so with nature itself. According to Kropotkin, understanding the role of mutual aid was completely dependent on time spent in the field, observing the behavior of animals in their natural condition.

In a letter quoted by Todes, Kropotkin described the proper direction of evolutionary theory as he saw it, returning to Darwin's original intent:

This is a theory of evolution which, following Bacon, recognized the importance of Mutual Aid—that is, of the social instinct—for the preservation of the species, and which, with Bacon, saw in it the primordial element of Ethics. . . . This is above all a return to the Darwinism which saw in Evolution a spontaneous

> result of the forces of Nature, and not, as Weismann and his disciples wished, an evolution predetermined (by the mechanisms of the Universe) by means of a substance possessed of an "immortal" soul—this Hegelian creation of Weismann, his germ plasm. A theory of Evolution finally, [which describes] a physiological evolution of organs [caused] by the new functions which they perform as the organism is placed in new conditions of existence.[29]

The fate of Kropotkin's theory, at least in the West, was alluded to earlier in this chapter. Nevertheless, the main ideas presented in the work discussed, consistent with the Russian naturalist tradition, continued to have an impact there. This is the argument of both Mark Adams and Daniel Todes, who have done extensive research in this area. Adams has argued that the Russian population geneticists bridged the gap between the naturalist and experimentalist traditions that had so concerned Kropotkin: "It is significant then that the Russian School is one of the earliest to draw from both traditions in order to clarify the evolutionary process. . . . And by turning the techniques of genetics onto the problems of evolution in a natural setting, Chetverikov did much to heal the unfortunate gap—in effect, by creating experimental population genetics and making evolutionary theory experimental."[30] In the same article, Adams points up another of the concerns that struck Kropotkin and was later emphasized by the Russian geneticists, that is, "that the experimental work under Chetverikov's direction, on a naturally occurring *Drosophila* population, led to the development of clear ideas concerning the influence of genetic and *environmental* backgrounds on the fitnesses and effects of genes."[31] I cite these examples not as direct results of Kropotkin's work but rather as indicative of the importance of the environmental context within which these biologists worked, and to provide a transition to the work of the next character in this story.

Theoretical Context and Advance

My review of the development of Darwin's theory around the turn of the century in these first two chapters highlights the difficulty naturalists faced regarding the mechanism of natural selection acting on groups. The issues of the nature of the individual and the evolutionary explanation of cooperative behavior were particularly challenging. Indeed, in the case of

Kropotkin, there is ambiguity in his ideas. He emphasized the role of the environment in shaping the interactions between members of a group that provided the basis for his theory of mutual aid. Simultaneously, he rejected Weismann's argument against the inheritance of acquired characteristics caused by changes in the environment. These are two distinct points that Kropotkin consistently confused. The first addresses the behavior of individuals in groups, and the second concerns the mechanism of inheritance between individuals and their offspring.

For the naturalists working during this period, ultimately the problem was to understand the transmission and fitness of certain group behaviors. As we have seen, the morphology and behavior of the social insects was more susceptible from the germplasm perspective, where the environment appears to play only a contingent role. On the other hand, the social behavior of the "higher" vertebrates allowed for more focus on the interaction of the environment and the development of social behavior.[32]

The interest in hereditary transmission and the evolution of social behavior only increased in the early twentieth century. At Oxford in the 1920s these were the concerns of E. B. Ford, Julian Huxley, and Charles Elton and they were to become the focus of their student, Vero Copner Wynne-Edwards.

Indeed, the intellectual ecology at Oxford was to play a significant role in the development of group selection theory. Immediately following World War I, Oxford was undergoing a major transition from an institution renowned for its devotion to the classical humanities to one in which scientific research became equally valued. As historian Jack Morrell has intimately detailed, the rise of the sciences at Oxford resulted from the sustained efforts of a number of eminent scientists and administrators, and these efforts were particularly successful in zoology, which he described as "widely recognized as an enclave of excellence."[33] In the Linacre department, under the influence of the comparative anatomist Edwin S. Goodrich, who held the Linacre chair for twenty-four years from 1921 to 1943, and Julian Huxley, a Balliol graduate who spent six postwar years at Oxford, from 1919 to 1925, they maintained an atmosphere that encouraged wide-ranging research in a Darwinian framework. As chair, Goodrich appointed star graduates from his own department as demonstrators (including Gavin De Beer, Charles Elton, E. B. Ford, and J. Z. Young). Huxley, whose interests ranged across behavior, genetics, embryology, and ecology, fostered an approach that moved biology from the purely laboratory-based methods of descriptive morphology and classification into the experimental and field-based mode that produced evolutionary and ecological accounts of living things in their natural environment.[34]

In the Hope department of entomology, the adherence to the Darwinian perspective was perhaps even more pervasive. The Hope chair was occupied from 1893 to 1932 by Edward Poulton, whom Jack Morrell described as "an ardent Darwinian" and "a charitable man, [who] could forgive anything save disbelief in Darwinian evolution."[35] Poulton's contributions to the study of evolution at Oxford included the support he provided to E. B. Ford in his development of ecological genetics with R. A. Fisher and an early (1921) push for experimental work on melanic moths. He argued that "Mendelism was a very valuable reinforcement of Darwin's theory of evolution by natural selection and urged that more experiments be done on the hereditary transmission of small variations."[36] Poulton's influence persisted in his successor to the Hope chair, Geoffrey Carpenter, another Oxonian who was his friend, protégé, and collaborator. Carpenter continued the Darwinian selectionist tradition and cowrote a volume on mimicry with E. B. Ford.[37]

The rise of biology at Oxford was further enhanced by the establishment of the Bureau of Animal Population in 1932 under the direction of Charles Elton and the Edward Grey Institute of Field Ornithology in 1938, which was initially led by W. B. Alexander but flourished under the direction of David Lack, who was appointed after the war. The influence of these two entities, and especially of their directors, on the development of group selection theory will be elaborated in chapters 3 and 6. Though Bowler's thesis regarding the eclipse of Darwinism provides an apt description of the broader Anglo-American intellectual environment, the brief account here demonstrates that the eclipse was apparently not visible from the environs of Oxford University.

CHAPTER THREE

Vero Copner Wynne-Edwards

The Early Years

Vero Copner Wynne-Edwards was born in 1906, the son of John Roslindale Wynne-Edwards, a canon in the Church of England and headmaster of Leeds Grammar School. The fifth of six children, Wynne, as he was called by family and friends, attended Leeds Grammar School and Rugby School, where he developed his early interest in astronomy and natural history. In his initial astronomical pursuits, Wynne used an uncle's two-and-a-half-inch telescope to record daily sunspot configurations. He described his amateur astronomer's notebook in his 1985 memoir: "The notebook also had diagrams of the day-to-day positions of Jupiter's visible satellites, and of the slowly changing array of the four bright planets, all visible in that winter evening sky. In the days when streets were lit with gas-mantle lamps, the stars shone bright on clear nights even in the city sky."[1]

During the same period, young Wynne also kept a natural history diary in which he recorded daily temperatures and rainfall, maintained a bird log, and occasionally wrote short descriptions of the behaviors and interactions of animals he had observed. The passage above not only attests to his youthful meticulousness and enthusiasm but also hints at the myriad changes that had occurred since his first experiences with natural history. Wynne-Edwards's early botanizing in the English countryside

FIGURE 5. V. C. Wynne-Edwards, Culterty Field Station. David Jenkins and Adam Watson, "Obituary: Vero Copner Wynne-Edwards (1906-1997)," *Ibis* 139, no. 2 (1997): 415. Reproduced with permission of Blackwell Publishing Ltd.

may not have differed significantly from that of the young Charles Darwin. Nevertheless, as his education and later career progressed, he would come to face the challenges of the development of twentieth-century life sciences—a more sophisticated understanding of Darwin's theory in relation to nature.

Expanding Horizons

In 1920, at age thirteen, Wynne-Edwards went on to Rugby. He had graduated from his uncle's two-and-a-half-inch telescope to a nine-inch equatorial model. He now had at his disposal a natural history museum and a burgeoning natural history society. While at Rugby, the young Wynne-Edwards was much impressed by some visiting lecturers: "I remember especially Sir Ernest Shackleton, on the eve of his last Antarctic expedition in the *Quest*, on which he was to die in January of 1922. . . . Another visitor who lectured 'awfully well' was Julian Huxley of whom I was later to see more."[2]

The lecture Huxley gave was on the 1921 Oxford expedition to Spitsbergen, the Arctic island off the coast of Norway. The expedition, under the leadership of F. C. R. Jourdain, vice president of the British Ornithological Union, was organized for two main parties. The first party, under the direction of Huxley, was to investigate the migratory movements and breeding habits of rare Arctic species. The second was to provide a geographical and geological study of the northeast region of the main island. The members of the party who were to be influential to Wynne-Edwards included not only Huxley but also his assistant, Charles Elton, as well as Alexander Carr-Saunders. All three of these expedition members became significant in various ways, which I will discuss later in this chapter and in chapter 5.

In Wynne's account of another of the Natural History Society lectures, one can see the legacy of William Paley that so influenced naturalists of Darwin's generation. The lecture, on stars, was presented by a Father Coortie of Stoneyhurst. The diary entry goes on for five pages recounting the startling size and shape of the heavens, which greatly impressed young Wynne. Father Coortie also discussed the ways technology was increasing our knowledge of the universe. He listed the largest telescopes under British dominion, including "Victoria" in British Columbia, at sixty inches. He also described the largest telescope in the world, at the Mount Wilson Observatory in California, which had a seventy-two-inch reflector. These great telescopes, in conjunction with the use of spectroscopy and photography, had given astronomers new insights into the true nature of the heavens, including the size, shape, composition, and origin of the stars. Despite the important contributions of modern technology, the ultimate lesson of the lecture was unmistakably vintage nineteenth century. "Father Coortie showed us some photographs of the moon at the close of the lecture. He finished his lecture with a synopsis, in which he impressed on us that there is an arrangement and pattern among the stars. Where there is a pattern, there is a designer. The stars are the greatest work of the great Designer. Good day. Fine night."[3]

It is fascinating to note the continued influence of the natural theological argument, which had been most convincingly presented over one hundred years earlier in Paley's *Natural Theology* and later expanded in the *Bridgewater Treatises*.[4] Some of Wynne-Edwards's later writings, which appealed to concepts such as the "balance of nature" or "optimum number" for a particular population, struck many of his critics as a return to this kind of outmoded natural theological thinking. These criticisms, among others, will be further examined and discussed in chapter 6.

In 1921 Wynne's father retired from Leeds and became rector of the parish of Kirklington, thirty miles to the north. The family house at Austwick was let, and owning a car made more far-ranging botanizing and natural history field trips possible. Wynne continued to hone his field skills and his eye for detail. On a hike up Ingleborough, a nearby peak, he discovered sandwort, a low-growing plant with white flowers, about one centimeter across, that was easily overlooked because two similar varieties grew in the same communities. On further research, he determined that the plant was *Arenaria gothica*, first discovered in the area only ten years earlier. At that time only one British handbook included the species. He requested it for his birthday.[5] Wynne-Edwards left Rugby in 1924 with hopes of taking part in Himalayan expeditions studying alpine fauna and flora. The headmaster and his father, perhaps not surprisingly, had more practical plans. The headmaster suggested Wynne study medicine so he could act as an expedition doctor, and his father suggested further formal study. In his memoir he reflects on the breadth of his education. "I often looked back in astonishment at the variety of things I had learned there—from Milton's poetry to military proficiency, and from theology to thermodynamics."[6]

In the end, Wynne entered New College, Oxford, in October 1924 to read zoology, with Julian Huxley as his tutor.

Oxford

The move to Oxford was fortunate for Wynne-Edwards, since the quality of the instructors he encountered there was unmatched by any other university at the time. After Wynne's first year, Huxley accepted the chair of zoology at King's College and moved to London in December 1925. Huxley's successor as tutor was Charles Elton (a former student of Huxley's), whose influence, according to Wynne-Edwards, was to be much more specific and enduring.[7] Elton, a pioneer of animal ecology who founded the Bureau of Animal Population at Oxford University in 1932, is credited with having sparked, from their first tutorial, an interest in population ecology that fueled the rest of Wynne-Edwards's intellectual life. Elton's lasting influence is perhaps not surprising, given some of his commonalties with Wynne-Edwards. As an undergraduate at Oxford, Elton had developed his own original ideas about the processes of natural selection by observing animals in the field. Like Wynne-Edwards, he was deeply influenced by the notion that animals do not "sit about waiting for the environment mindlessly to select the fittest to survive, as plants must. They generally can and often do move from places in which they are not doing well into places in which they may do better."[8] The recognition of the organism's active role in evolution was echoed by Wynne-Edwards as he developed his theory of group selection. Elton also shared the experience of being a naturalist in the Arctic. After the Spitsbergen expedition in 1921 he participated in the Merton College (Oxford) expedition of 1923, which attempted to circumnavigate North East Land, and he was in charge of scientific work on the Oxford University Arctic expedition of 1924.[9] In a 1953 letter to Wynne-Edwards he wrote: "It is very cheering to hear from one's friends, and I am glad if animal ecology begins to be generally treated as a serious science. I still feel rather a naturalist with scientific leanings, but don't say I said so."[10] Wynne-Edwards also credits Charles Elton for introducing him to the work of Alexander Carr-Saunders. Carr-Saunders (another of the Spitsbergen members) was trained in zoology at Oxford and later was a demonstrator in zoology there. While at Oxford, in 1922 he published *The Population Problem,* which anticipated some aspects of Wynne-Edwards's group selection theory by forty years. Carr-Saunders left Oxford the year before Wynne-Edwards matriculated, and he later became director of the London School of Economics.[11] Although Wynne-Edwards found Carr-Saunders's book immediately intriguing, it would not be until much later, while working on a chapter of his magnum opus in 1959, that he realized how important and influential it was.

The other lecturers of the time were also standouts in their respective fields. Gavin De Beer gave the lectures in embryology and experimental zoology, and E. B. Ford conducted the genetics courses. Both of these instructors became members of the Royal Society, and Gavin De Beer was appointed director of the Natural History Museum at South Kensington and later knighted for his contributions to science. The zoology department that had been assembled by Edwin S. Goodrich, the Linacre chair of zoology, was the success story among the small departments at Oxford, which had few college fellows or none. After losing two of his senior men in World War I, Goodrich put together his staff exclusively from Oxford graduates taught by himself and Julian Huxley. In his chapter on the nonmedical sciences in *The History of the University of Oxford*, Jack Morrell succinctly describes the development of the department that would shape Wynne-Edwards's approach to nature.

> Goodrich provided the basic morphological training but did not force his staff into his own research mould: the nearest to him was [J. Z.] Young, who added his interest in nerves to Goodrich's emphasis on structure. Huxley inspired his students with his ideas about the importance of fieldwork, animal behaviour, experimental embryology, and population dynamics. Through Huxley and Elton, Oxford Zoology became strongly ecological. Goodrich and Huxley were staunch Darwinians, and the department pursued a diversity of work within a Darwinian framework: witness Goodrich's Festschrift produced by Oxford colleagues and pupils with essays on sexual selection by Huxley, ecological genetics by Ford, experimental embryology by De Beer, human evolution by A. M. Carr-Saunders, animal ecology by Elton, breeding seasons by Baker, the nervous system by Young, dermal bones by J. A. Thomas-Moy and bird evolution by Tucker—all topics on which these past or present demonstrators in Goodrich's department were leading experts. Not surprisingly, pupil-teacher lineages were established over a short time: Huxley taught De Beer, who taught Young, who taught P. B. Medawar, who was to win the Nobel Prize in 1960.[12]

Morrell had corresponded with Wynne-Edwards while conducting the research for this chapter. In his reply to Morrell's inquiry, Wynne-Edwards included some personal details that did not become part of the final chapter. First, he pointed out that although Huxley's presence was a great motivator in his decision to attend Oxford, family history also played a role; Wynne's father, grandfather, and eldest brother Robert had all attended Oxford. Second, he credited Huxley with creating the opportunity for him to meet the woman who became his life partner—Jeannie Morris, another Oxford student. In his 1987 letter he wrote: "Though I was never formally taught by him in the Zoology department, it is worth mentioning that my wife, who also came to Oxford in 1924, intending to read Botany, was won over to Zoology (which she had not studied at school) by Julian's first year lectures. But for that, she and I might never have met."[13]

After discussing the contributions of his former professors and colleagues, he concluded his letter by noting, "Now, nearly 60 years later we can look back on Oxford University as one of the great fountainheads from which innovations have sprung in the fields of vertebrate ecology and ethology, population dynamics, sociobiology and evolutionary theory, currently involving thousands of research workers worldwide. Dr. David

Lack, the first director of the EGI [Edward Grey Institute], was himself one of the foremost pioneers."[14]

Obviously, Wynne-Edwards's recollections of his time at Oxford are positive. His diaries record various field trips: the summer of 1925 spent working at a hatchery in Port St. Mary and aboard the fishing boat *Manx Princess,* studying various fishes and shrimp; and multiple excursions to the Bagley Wood with Elton to capture and study voles, as well as commentaries on various lectures. In January 1926, Sir Arthur Conan Doyle lectured to an audience of nearly one thousand. The lecture was the source of an interesting entry in Wynne's daily diary:

> Conan Doyle gave a lecture in the evening to the Junior Scientists on his psychic researches. . . . It is interesting to see people laugh at him as they laughed at Galileo and Darwin, though I do not suggest that he is as great a man as they. In five and twenty years all that he says must be confirmed, though perhaps somewhat modified. He admitted that only the beginnings had been made. He is a clever speaker, though not always very distinct, being hampered by his mustache wh. seems to get into his mouth. He produced plenty of sound evidence, and as the "skipper" said in his presidential vote of thanks, "he is surely a bold man who will refuse altogether to believe in the truth of what Sir Arthur has told us." I believe certainly that there is no fake about it, though I do not agree with him about the nature of the spirits. Some of his remarks were ludicrous. (Faith grows dim with time! 7.iii.26.)[15]

The parenthetical remark, added a couple of months after the lecture according to the date provided, may be the most interesting part of the passage, and the entry bears scrutiny for two reasons. First, it reflects Wynne's response to unorthodoxy, a response that one familiar with his later work might expect. He is skeptical but willing to reserve judgment until the data are in and the theory has had a chance to be carefully tested. Second, the parenthetical comment illustrates Wynne-Edwards's breaking from the faith-based natural theological approach to understanding nature. Like many of the students of his generation, the more he studied and understood Darwin's theory of evolution by natural selection, the more powerful it appeared.

Wynne-Edwards graduated with first-class honors in zoology in 1927 and became senior scholar of New College for 1927–29. He was also appointed to a student probationership at Plymouth Marine Biological Laboratory. His earliest work was in the field of marine zoology; simultaneously, he published several papers on the wintering behavior of starlings. In these articles the earliest strains of Wynne-Edwards's thoughts on dispersion and breeding begin to surface. An article published in *British Birds*, "The Behaviour of Starlings in Winter," for example, is generally descriptive, but the discussions of species-level selection and nonbreeding behavior (especially in relation to the rapidly increasing population) mark them as topics of particular theoretical interest to him. He rejected the explanation provided by a preceding study that the growth of the starling population followed the advance of agriculture in the region, which increased the availability of food. This early rejection of food supply as the limiting factor in population level foreshadows the later intense debates with David Lack. He argued that the agricultural development in the southwest of England had occurred over a much longer period than the fifty years that marked the population and range expansion of the starlings.[16]

In another challenge to contemporary thinking, Wynne-Edwards regarded both the "safety in numbers" and the "lack of suitable sites" explanations of the roosting habit as suggestive but not convincing. He argued that the roost provides no protection at night, since the birds can easily be plucked from their perches as they sleep with nary a rustling from their roostmates. And he pointed out that a hundred birds would be just as effective (or ineffective, as the case may be) as the thousands that typically occupy a roost. Regarding the availability of roost sites, he pointed out that landowners who destroyed roosts in an attempt to get rid of the starlings generally found the group reestablished at a nearby site the next day.[17] Wynne-Edwards's rejection of the contemporary explanations indicates how important understanding population regulation and its connection to behavior was to him from the very beginning. From the early field trips under the supervision of Charles Elton through his first published scientific papers, questions regarding the maintenance of population levels and the effects of population fluctuations on individuals and groups formed the basis of Wynne-Edwards's approach to the study of nature.

The North Atlantic and Arctic

In 1930 Wynne-Edwards was offered an assistant professorship at McGill University in Montreal. This appointment presented one of those providential opportunities that are often so important in a developing scientific career. His transatlantic observations led to the publication in 1935 of a paper in the *Proceedings of the Boston Natural History Society*, "On the Habits and Distribution of Birds on the North Atlantic." This earned him the society's Walker Prize, the first major prize of his young career.

Wynne-Edwards got his early inspiration for the project leading to this award on the initial voyage to McGill, aboard the Canadian Pacific Liner *Empress of Scotland*, which left Southampton in September 1930.[18] His ornithological observations on this trip led him to develop an outline of the basic pattern of inshore (coastal), offshore (out to the edge of the continental shelf), and pelagic (deepwater) zones of seabird distribution. On his arrival at McGill, Wynne-Edwards wrote to Sir Edward Beatty, president of the Canadian Pacific Line and a chancellor of McGill University, asking if he could be given free passage to conduct bird studies. Unfortunately, Beatty viewed his eagerness as importunate, and the Canadian Pacific option was summarily closed. Luckily for Wynne-Edwards, the university principal, Sir Arthur Currie, directed him to the dean of the medical faculty at McGill, whose brother-in-law was the sole Canadian director on the board of Cunard Lines, the only other line operating between Montreal and England. In short, the project was approved and Wynne-Edwards made four round-trip voyages in May, June, August, and September of 1933 studying the distribution of the avifauna. This path across the North Atlantic is an important aspect of the development of Wynne-Edwards's thoughts about nature and animal communities.

With this background of experience, Wynne-Edwards opened the first section of the 1935 paper with the following vivid passage:

> Nowhere on land, even in the Sahara, the prairies or the steppes of Asia, can such a vast expanse of monotony be found as on the great oceans. Because of the relatively small temperature variation at any point and the unending circulation of their waters, they present a uniformity of conditions unparalleled elsewhere on this earth. Yet it is hardly necessary to state that in spite of this prevailing sameness not all the birds primarily adapted to obtain their livelihood from the sea, even in a restricted area like the North Atlantic, belong to a single ecological community. Fulmars and cormorants, for example, might pass their whole lives without seeing one another, and could only do so at special times and places, for they belong to two communities as distinct as those of forest and fen, and their paths seldom cross. The factors which differentiate one community from another are not by any means understood, but present problems of no small interest.[19]

This passage is significant because it illustrates the type of environment where Wynne-Edwards conducted his formative fieldwork. Exposure to these environmental conditions contributed to the development of his thinking regarding the "struggle for existence" and Darwinian selection. Darwin's own explanation of the concept of struggle from the *Origin* is important for comparison: "I use this term in a large and metaphorical sense including dependence of one being on another, and including (which is more important) not only the life of the individual, but the success in leaving progeny. Two canine animals, in a time of dearth, may truly be said to struggle with each other over which shall get food and live. But a plant on the edge of a desert is said to struggle for life against the drought."[20] Although the emphasis of Darwin's idea of struggle was on intraspecific competition, this broader construal of struggle had long been recognized (as I discussed in chapter 2). This interpretation is also closely connected, conceptually if not historically, to the impressions of Kropotkin discussed in chapter 2. The following passages exemplify a common view of nature. The first is Kropotkin in the opening paragraph of the introduction to *Mutual Aid*:

> Two aspects of animal life impressed me most during the journeys which I made in my youth in Eastern Siberia and Northern Manchuria. One of them was the extreme severity of the struggle for existence which most species of animals have to carry on against an inclement Nature; the enormous destruction of life which periodically results from natural agencies; and the consequent paucity of life over the vast territory which fell under my observation. And the other was, that even in those few spots where animal life teemed in abundance, I failed to find—although I was eagerly looking for it—that bitter struggle for the means of existence among animals belonging to the same species, which was considered by most Darwinists (though not always by Darwin himself) as the dominant characteristic of struggle for life, and the main factor of evolution.[21]

This opening is strikingly similar in its description of nature to the introduction of Wynne-Edwards's 1935 article, and Kropotkin's second point in the passage above becomes the calling card of Wynne-Edwards's later work. In a 1952 article in the *Auk,* on the zoology of the Baird expedition, Wynne-Edwards's characterization of the struggle is even more explicitly similar to Kropotkin's, although the latter is not cited:

> Except perhaps among carnivorous predators, competition between individuals for space and nourishment seems commonly to be reduced to a low level among members of the arctic flora and fauna; they live somewhat like weeds, the secret of whose success lies in their ability to exploit transient conditions while they last, in the absence of serious competition. In the Arctic the struggle for existence is overwhelmingly against the physical world, now sufficiently benign, now below the threshold for successful reproduction, and now so violent that life is swept away, after which recolonization alone can restore it.[22]

Here again, a view of nature is distinctly described that differed significantly from the tropic environs that so influenced Darwin and his codiscoverer Alfred Russel Wallace. While a member of the faculty at McGill, Wynne-Edwards participated in a number of scientific expeditions, all in similar northern climes.

An expedition to northern Labrador produced some of the first clear indications of his later work on group selection. In the summer of 1937 Wynne-Edwards accompanied Commander Donald B. MacMillan and a crew of young science students on an expedition to Frobisher Bay in the Gloucester schooner *Gertrude Thebaud*. Wynne-Edwards was part of the four-member scientific crew, which also included a geologist, a botanist, and another general zoologist. The expedition was to last from July 16, 1937, to the end of August. The purpose of the expedition was to map the coastline of Baffin Island as well as to collect scientific specimens and observations. Here again, Wynne-Edwards's diaries paint a picture of unforgiving nature in a sparsely populated land. In his entry from the July 29, a day that began with a climb up a nearly three-thousand-foot peak and ended with the *Thebaud* run aground and heeled over 30.5 degrees to port, Wynne-Edwards wrote,

> All around was evidence of very recent decrease in the glaciation. Our empty valley showed a "high-water" mark in the uppermost corrie, below which the rocks were pale clay-colored, without a blade of vegetation, even moss. This was 100' at least above the "glacierette" which now occupies the corrie. At the edge of the snowfield are roches mountainées which look as fresh as if they had been exposed for the first time this spring. They are cracked into oblong blocks as regular as building stone, with a smooth curved surface.[23]

The entry goes on to recount the contribution to geographical knowledge he had made by establishing the height of the peak at 2,780' ± 25', and

FIGURE 6. The *Gertrude Thebaud* heeled over in Forbisher Bay, 1937. Photo courtesy of Captain Jim Sharp.

his constructing, later that evening, a small hutch for the pair of lemmings he had collected. It was only four hours later that Wynne was awakened because the ship had run aground. They spent the morning abandoning ship and ferrying all the necessary gear to shore. As the tide came in and the crew manned the pumps, the ship righted itself and was saved. Once the voyage was safely under way again, Wynne-Edwards resumed his daily routine of sketching a coastal map in the morning and then going ashore to collect during the day. In the evenings he pressed his botanical samples, skinned the various birds and small mammals he had collected, and wrote his entry for the day.[24] Many of Wynne-Edwards's specimens were donated to various Canadian museums, and his collection and observations of fulmars (*Fulmarus glacialis*) led to an important transitional article on their breeding behavior.

This paper, titled "Intermittent Breeding of the Fulmar, with Some General Observations on Non-breeding in Sea-Birds" and published in the *Proceedings of the Zoological Society of London* (1939), marked a transition from purely observational, descriptive, nineteenth-century natural history to an approach more concerned with theory. At the same time, Wynne-Edwards laid the groundwork for later investigation of non-breeding behavior as a mechanism for population regulation. He wrote that based on previous observations (his own as well as others') it was estimated that only between one-third and two-fifths of the fulmars present at a particular breeding colony appeared to be engaged in reproduction. This surprising fact required explanation. If, consistent with Darwinian theory, individuals are constantly striving to increase their representation in subsequent generations, why were so many of these fulmars not trying to reproduce? In the next chapter I will examine more closely how Wynne-Edwards problematized nonbreeding behavior and how it led to the development of his theory of group selection.

Over the next decade Wynne-Edwards would participate in several expeditions in the northern reaches of Canada. Each of these trips, according the notes in his diaries as well as the published accounts, reinforced his view of nature. His Arctic nature, like Kropotkin's, was often—though certainly not always—sparsely populated, and the behavior of its inhabitants continued to intrigue him because it lacked the "Darwinian" ferocity of struggle and competition. In the summers of 1944 and 1945, the Fishery Research Board of Canada sent Wynne-Edwards and Dr. Ronald Grant on two expeditions to report on the fish resources of the Mackenzie and Yukon rivers. This research was part of a reconnaissance commissioned by the federal government of the resources of the Far Northwest and its potential for settling men discharged from the armed forces after the war. In 1944 the pair canoed 1,300 miles down the Slave-Mackenzie system in a freight canoe, from Fort Smith to Aklavik on the Beaufort Sea. In 1945 they spent two months in the Yukon Territory, traveling by truck on the new Alaska Highway and its feeder roads and descending 400 miles down the Lewes-Yukon rivers, from Whitehorse to Dawson City, in a twenty-seven-foot flat-bottomed boat. Most of the observations recorded on these trips concerned size of catch and productivity levels during the various seasons. Nevertheless, Wynne-Edwards continued to be struck by the unusual—that is to say, non-Darwinian—character of these surroundings.

> The Mackenzie runs for the most part over vast alluvial plains, often in several channels separated by willow-covered islands. One can seldom see further than the forested shores—often more than a mile apart—on either side, and the length of the reach ahead and the reach astern. It carries a vast volume of water, warmed in summer by the hot Alberta sunshine, down to the Arctic Sea, and when it gets there three or four weeks later it is still much warmer than the innumerable ponds in the permafrost muskeg on either side. We were the first

> biologists to interest ourselves in the small kinds of fish as well as the large, and we found that several temperate-zone species extended all the way down to the delta, thus enlarging their previously known ranges by more than a thousand miles. Food productivity of the river is remarkably high.
>
> The upper Yukon is entirely different. Many of the headwaters are in snowy mountains; the rate of descent is much faster and the speed of flow about 6 knots, exceptional for a great river in a mature bed. The cold water swirls and boils as it goes. Not far above Dawson City, it receives from a tributary an enormous load of white pumice in suspension, and the mixed water looks like café au lait and makes a curious low hissing sound. The scenery is majestic and I have little doubt the Yukon is one of the world's most beautiful rivers. King Salmon migrate up from the Bering Sea, some of them reaching 1750 miles from the mouth; and because of the current against them, even though they hug the banks, it presumably means swimming to two or three times that distance through the water. They eat nothing en route. It seems an almost incredible mechanical achievement on a kilogram or two of stored fat.[25]

In 1950 Wynne-Edwards returned to the great white North for his penultimate expedition. He and Alexander Anderson composed the zoological unit of the expedition, under the command of Colonel Patrick Baird and funded by the Arctic Institute of North America. Colonel Baird wrote in his account of the expedition that the large scientific party was warranted because central Baffin Island was sufficiently unknown to require the members' mixed scientific expertise. One of the first scientific questions to be addressed was how the ice cap (ninety miles long and forty miles wide) maintains itself. It was previously thought that the cap was maintained as seasonal snow became permanent. However, the summer's observation provided evidence that it was the adfreezing of the meltwater as each winter's snow cover melted. Meanwhile, Wynne-Edwards spent the summer collecting the six passerine species nesting there and studying their breeding biology.[26] He was to return to Baffin for a final expedition in 1953. As will become evident in the following chapters, these experiences informed his views of nature in a way that greatly influenced the development of his theory. Similar to Kropotkin's experience, and contrary to that of Darwin and his neo-Darwinian foil David Lack, Wynne-Edwards's direct experience of the Arctic environment him led him to a significantly different conception of the role of struggle in nature. His perspective emphasized the social organization and cooperative effort necessary for a group of organisms to survive in the harsh Arctic environment.

Aberdeen

The final move of Wynne-Edwards's career occurred in 1945 after he served in the Canadian Naval Reserve in World War II. During the war he remained at McGill teaching electronic physics to radar mechanics in the Royal Canadian Air Force. The summer after the war, while on his trip in the Yukon, he received a letter from his wife reporting that the Regius chair of natural history at the University of Aberdeen had become vacant. Wynne-Edwards applied for the position and was hired later that year. He served as Regius chair until 1974 and presided over the establishment of the Culterty Field Station for research in ecology; he also was a leader in various scientific, environmental, and ecological associations. It was at Aberdeen that Wynne-Edwards's interest in population structure and breeding behavior would develop into a full-blown theory of group selection. The development of this theory will be the subject of chapters 4 and 5.

The first set of articles reviewed in this chapter, representing Wynne-Edwards's early professional work, may have been of interest to a small collection of naturalists concerned with the lives of birds; but with the advent of the modern synthesis Wynne-Edwards recognized a fundamental shift in the mode of thought about evolution that was perhaps more in line with his own interests. In November 1948 he gave a paper to the Oxford Ornithological Society, "The Nature of Subspecies," in which he discussed the importance of the shift. In his introductory remarks he cited the work of E. B. Ford on butterflies, as well as Dobzhansky's, Mayr's, and Huxley's core contributions to the development of the modern synthesis. He wrote, "The fundamental new idea is that populations, rather than independent individuals, are the basic units upon which evolutionary processes act."[27] The acknowledgment of this shift, and the 1954 publication of David Lack's *Natural Regulation of Animal Numbers*, led him to begin the focus on population and in particular population regulation that shaped the rest of his professional life. Wynne-Edwards's critical review of Lack's book, and a 1955 article that he began by quoting Lack's assertion that birds reproduce as rapidly as they can, mark the beginning of his quest to explain population structure through social behavior. His theory that social

FIGURE 7. Marischal College, Aberdeen University. Photochrome Collection, vol. 2, U.S. Historical Archive, St. Augustine, Florida.

behavior had evolved through the mechanism of group selection in order to maintain populations at the proper level in relation to their food supply would bring him the attention of the entire biological community and would set in motion a debate over the role of group selection that persists into the present.

CHAPTER FOUR

Theory Development

Counterexperience: Lack in the Galápagos

While Wynne-Edwards was taking notes on the breeding of fulmars at Cape Searle, Julian Huxley was encouraging a young friend, David Lack, to make a trip to the tropics to compare territorial behavior in a group of related species. Initially Lack considered African weavers as the subject of this proposed study, but on reading P. R. Lowe's 1936 *Ibis* paper on the Galápagos finches, he decided they would be most appropriate. Lack's experience on the Galápagos was not idyllic; nevertheless, it provided the core data for one of his most important contributions to biology—his work on the adaptations of closely related species.

On returning from the Galápagos with a collection of live finches, Lack realized that the birds were in molt and were not likely to survive the trip to England. Therefore he stopped in California to nurse the birds and study the collection of skins at the California Academy of Sciences. He stayed in the United States for five months, studying in California and also visiting Ernst Mayr, at the Museum of Natural History in New York, with whom he had many influential conversations regarding systematics and speciation. Mayr was to remain a lifelong friend and a significant influence.

Lack studied the finches because he originally planned to write a book for sixth-form pupils demonstrating evolutionary principles, using the Galápagos finches as examples; the final product was significantly more.

FIGURE 8. David Lack in the Galápagos, 1938. David Lack Papers, Alexander Library, Oxford University, box 3, file 37. Photo courtesy of Peter Lack.

In his memoir, Lack pointed out that his position—that the differences in beak morphology were adaptive—ran counter to the standard view. Before the publication of his book, almost all subspecific differences in animals were thought to be nonadaptive. The only exceptions were size differences and coloration of certain desert larks, and specific recognition characters that reduced hybridization. Lack concluded his account of the importance of his realization about the finches with these lines:

> I reached my conclusion, that most subspecific and specific differences in Darwin's finches are adaptive, and that ecological isolation is essential for the persistence of new species, only when reconsidering my observations five years

> after I was in the Galapagos. In this, of course, I had a distinguished precedent, as Charles Darwin did not perceive the evolutionary significance of the finches until several years after his visit (partly because it was first necessary for Gould to tell him that they were a group found solely in the Galapagos).[1]

Lack's reassessment of the finches was also heavily influenced by Huxley's recently published *Evolution: The Modern Synthesis,* and even more by Mayr's *Systematics and the Origin of Species*. One of the central questions of the time was how populations became isolated as a prelude to speciation. Huxley's approach was broad, including geographical, ecological, and genetic isolating mechanisms. He thought that ecological divergence would occur when there was little or no competition but different parts of a population were under different selection pressures, causing them to adjust to different modes of life. Mayr's position was that only geographical isolation could lead to speciation. From his point of view ecological differences were merely the result of spatial separation. Initially, Lack made aspects of both these positions part of his own viewpoint, but ultimately he came to accept Mayr's position as more correct. There was one more influence on Lack's thinking that proved decisive in his interpretation of the finch data: Georgii Gause's principle of competitive exclusion, which states that two species living together cannot occupy the same ecological niche.[2] In the light of these ideas, Lack realized that the finches had originally diverged when they were geographically isolated. This led to the conclusion that, when they met again, competition was reduced as a result of the ecological specialization that had occurred, and the distinct species were now able to coexist. Lack could now assert that the differences between species, previously assumed to be unimportant, were actually adaptations and served as mechanisms to avoid competition.[3]

This brief survey of Lack's finch work is relevant for a couple of reasons. First, it provides a nice illustration of Lack's connection to the ideas that were fundamental to the developing modern synthesis, particularly the emphasis on adaptation and natural selection. Second, it gives an indication of the type of argument Lack used to support his theoretical claims. Sharon Kingsland made note of this style in her presentation of Lack's work.

> His argument, he realized, was based largely on inference and circumstantial evidence; he had not proven these inferences experimentally. Nor could he imagine an easy way to demonstrate the truth of his carefully constructed argu-

FIGURE 9. Wynne-Edwards, Baffin Island. Vero Wynne-Edwards Collection, Queen's University Archive, box 8. Photo courtesy of Queen's University Archive.

> ment, for competition among mobile creatures was hard to test. . . . Instead of direct proof, Lack relied on deduction from two likely premises: that divergence arose from geographical isolation alone and that species were limited by food supply. The facts which he brought to bear on his argument supported his interpretation, but it was not a clinching argument.[4]

According to Kingsland, Lack often struck a tone of strong advocacy to compensate for the circumstantial nature of his argument. This aspect of his approach is confirmed in Peter Crowcroft's work on the Bureau of Animal Population. He wrote of Lack at the Edward Grey Institute, "The more restricted scope of the work undertaken also made the place more susceptible to the dominance of its director, even if Lack had not remained rather a schoolmaster at heart. He had no interest or faith in the usefulness of experiments in elucidating ecological problems."[5] Lest one think Crowcroft's characterization of Lack ungenerous or merely the product of the strained relations that sometimes existed between their neighboring departments at Oxford, four of the six tributes to Lack, written by his friends and colleagues and accompanying his obituary, mentioned Lack's advocacy of single-factor explanations, an intolerance for alternative ideas, and too great a concern with controversy.[6]

Lack's connection to the modern synthesis was forged through his relationship with Huxley and Mayr and his increasing appreciation of the importance of adaptation and natural selection; nevertheless, the modern synthesis was not a monolith. For Wynne-Edwards, on the other hand, the modern synthesis opened the door to the analysis of populations and population regulation from an entirely different perspective. Perhaps it is not surprising that, although the modern synthesis has a name that is distinct and universally agreed on, it has no such agreed content or interpretation. Several conferences have been convened and many essays and monographs written to address this issue.[7] In the following section I will examine the views on higher-level selection of some of the major contributors to the modern synthesis.

The Modern Synthesis

During the modern synthesis, the approach to higher-level selection took an interesting turn as a result of the work of the population geneticists. The early mathematical models of Sewall Wright and the work of Theodosius Dobzhansky on flies were of particular significance. In his seminal work *Genetics and the Origin of Species*, Dobzhansky introduced the paradox of viability.[8] In a discussion of the level of variation a population must maintain to remain viable, Dobzhansky wrote, "Evolutionary plasticity can only be purchased at the ruthlessly dear price of continuously sacrificing individuals to death from unfavorable mutations."[9]

The evolutionary plasticity Dobzhansky invokes is a characteristic of a population. And the quotation above provides evidence that he was focusing on a group-level phenomenon and was interested in its importance to evolutionary theory. The increased variation of one population compared with that of a competing population conferred a fitness advantage on the more varied population even at a cost to the individuals making it up. These kinds of interpretations of the data produced by the population geneticists created the theoretical space within which Wynne-Edwards would pursue his theory. Dobzhansky also made interesting claims that the physiology of populations had been entirely neglected but was, at the same time, perhaps the most essential aspect of the theory of evolution. In his chapter on variation in natural populations he argued that, although the origin of variation was purely physiological, when injected into a population it entered the field of action of factors operating on a different level. According to Dobzhansky, "These factors, natural and artificial selection, the manner of breeding characteristic for the particular organism, its relation to the secular environment and to other organisms existing in the same medium, are ultimately, physiological, physical, and chemical, and yet their interactions obey rules *sui generis*, rules of the physiology of populations, not those of the physiology of individuals."[10]

His emphasis on higher-level selection, which he calls group selection (following Wright), was of obvious importance to Wynne-Edwards and has been previously underemphasized by authors writing about the synthesis. The contributions of Sewall Wright, whose earlier models influenced Dobzhansky's thinking, were also of interest and became increasingly so as he developed his ideas.[11] Sewall Wright's role will be more closely examined in chapter 6. This interest in population physiology was also a major interest of University of Chicago ecologist Thomas Park. Park's series of nine papers, "Studies on the Physiology of Populations," appeared between 1932 and 1939 and had a significant influence on the development of laboratory ecology. Park was consistent with Dobzhansky in that he too insisted that the physiology of populations was distinct from the physiology of individuals and could not be inferred from individual physiology.

The work of the other contributors to the modern synthesis, especially Ernst Mayr, George G. Simpson, and Julian Huxley, also addressed the possibility that the mechanism of natural selection functioned at a level above the individual.[12] This quotation from Ernst Mayr's *Systematics and the Origin of Species* provides a nice example of this thinking: "Darwin thought of individuals when he talked of competition, struggle for existence among variants, and survival of the fittest in a particular environment. Such a struggle among individuals leads to a gradual change of populations, but not to the origin of new groups. It is now being realized that species originate in general through the evolution of entire populations."[13]

FIGURE 10A. Theodosius Dobzhansky, Brazil, ca. 1965. Photo courtesy of the American Philosophical Society.
FIGURE 10B. Ernst Mayr, ca. 1955, neg. 334102. Photo courtesy of the library, American Museum of Natural History.

FIGURE 10C. Sewall Wright, ca. 1950. From *Biographical Memoirs* (Washington, DC: National Academies Press, 1994), 64:438. Copyright © 1994, National Academy of Sciences.

It is not difficult to imagine how this passage might have been interpreted by Lack and by Wynne-Edwards. From Lack's point of view, Mayr is merely arguing that new species originate through the aggregate action of a group of individuals constantly maximizing their individual fitness, whereas Wynne-Edwards would be most impressed by the final line of the quotation, emphasizing the evolution/selection of entire populations.

It is important to point out that this account relies solely on the first editions of the synthesis literature (Dobzhansky, Mayr, Simpson, and Huxley), and it has been argued that as the role of adaptation became increasingly emphasized across the biological spectrum, many of these broader interpretations of the action of natural selection were dropped from the later editions.[14] I point this out because, as I will demonstrate below, a similar reemphasis on the role of adaptation is fundamental to the criticism of group selection theory in the postsynthesis period.

Another interesting aspect of the modern synthesis period is that during this time there is a definitive shift among those considering group selection from the model organism approach to a more abstract mathematical approach. The work of the biologists of the early twentieth century, which focused on the social insects, was replaced by mathematical models of idealized populations. This was an unfortunate development for Wynne-Edwards.

Wynne-Edwards and Early Theory Development

As we have seen from his early work, to Wynne-Edwards the idea that animals reproduce as rapidly as possible was clearly off the mark. On January 24, 1955, he received a letter from his colleague C. M. Yonge, a member of the zoology department at the University of Glasgow. Yonge explained that the editor of *Discovery: A Monthly Popular Journal of Knowledge* was looking for someone to write an essay review of David Lack's new book: "I don't know how soon he would want this but doubtless you could do this as quickly as anyone with your intimate knowledge of the subject matter and the fact that you have a point of view which enables you to criticize constructively."[15] Wynne-Edwards replied immediately:

> I am prepared to have a shot at writing the review article for *Discovery*, though I know it would be difficult and it might turn out that I could not do it to my own satisfaction. I am up to the neck in the subject at present; my first paper is not

> yet published (though I am hoping to have proofs any day), and I am about to embark on a second. The trouble is that I think Lack has failed to penetrate the first principles of the subject, and in order to demonstrate this one must go very deep oneself: in fact, this is just what I am writing my two papers about.[16]

Six months later, the review had yet to be written. In a letter to William Dick, the editor of *Discovery*, Wynne-Edwards apologized and offered to let someone else write it. His own paper on population regulation (mentioned in the letter to Yonge) had received an unfavorable referee's report, and he was particularly disappointed: "I shall have to think again and try to present the material in a new and acceptable manner. . . . I feel pretty sure that some of Lack's fundamental concepts are mistaken, but in the present circumstances I should not like to say so in a review article, having so recently failed to convince a referee that this was the case."[17]

Ultimately, Wynne-Edwards wrote the review, which appeared in the October 1955 issue of *Discovery*. In it he praised the value of the broad-ranging work but took Lack to task for some of his hypothetical assertions. For example, he pointed out that according to Lack's view natality is the independent variable and mortality adjusts itself (through density-dependent effects) to match it.[18] He quoted Lack as asserting that "natural selection cannot favour smaller egg-number as such."[19] This is one of the particular hypothetical assertions Wynne-Edwards questioned. He countered that Lack had mistaken fecundity for fitness and that this was deeply problematic. If indeed the most fecund are not the most fit, then over time this stock will "very likely fall on evil days."[20] Further, according to Wynne-Edwards, selection could just as readily favor a lower reproductive rate as a higher one, differentially permitting the survival of those populations that continue to live in harmony with their environments.

The rest of the essay review presents some of Wynne-Edwards's own positions on population regulation and animal dispersion:

> It is no doubt a heritage of the Darwinian tradition that we tend to focus our attention on the struggle for existence, and the marvelous adaptations of animals for survival. . . . The struggle may sometimes be a desperate one, *unsuccessful even for the fittest*; but far off in the heart of its domain there may be places where, by contrast every desirable condition of life seems to be abundantly fulfilled. . . . The hardest struggle must obviously not be identified with the best world, nor with the surest chances of survival.[21]

Wynne-Edwards is again in line with Kropotkin's point of view that the fiercest struggle can produce a net loss for the stock, since even the fittest individuals are damaged as a result.

The invitation to review Lack's book was a result not only of Yonge's and Wynne-Edwards's long acquaintance but also of Yonge's awareness of the paper Wynne-Edwards had presented at the Eleventh International Ornithological Congress, held the previous summer in Basel, Switzerland. Wynne-Edwards attended the congress from May 29 to June 5, 1954, as the official delegate of the Royal Society of Edinburgh and gave a short paper titled "Low Reproductive Rates in Birds, Especially Sea-Birds." He had given drafts of the paper to various colleagues and was excited and apprehensive about their responses. Indeed, one of his potential readers, David Snow of the Edward Grey Institute of Field Ornithology at Oxford, complained in a letter from May 1954, "I'm sorry that your paper has been held up next door, and that I shall probably miss seeing it before the Congress. Elton's reactions to it should be of great interest."[22] The "next door" he referred to was the Bureau of Animal Population, which was in the same building as the Edward Grey Institute. The reader holding up the paper was none other than Charles Elton. His response, however, was not what Snow had anticipated. Elton wrote in a short letter to Wynne-Edwards that he found the paper interesting and was sure it would spur a great deal of discussion. His enthusiasm was particularly mild.

In his presentation to the congress, some of Wynne-Edwards's ideas regarding his developing theory of group selection were introduced in qualified terms. He wrote that "it is *theoretically possible* to regulate the numbers in the population by density dependent control of the recruitment rate alone" and that "control of this sort *could be* largely intrinsic, that is depending for its operation on the behavior-responses on the members of the population themselves." Later in the discussion section he suggested, "A collective response by a social group to general conditions of food productivity *does not appear so very much more abstract* and improbable than the corresponding individual responses by male birds in claiming a territory."[23] The concluding paragraph of this paper marks the first unequivocal statement of Wynne-Edwards's theory about the role of social behavior in limiting population.

> The theory that slowly-breeding birds have evolved a series of interrelated adaptations, giving them a great measure of autonomic control of their numbers,

> permits, at any rate, a rational explanation to be offered of many hitherto unconsidered or anomalous features of their breeding biology. It shows that if they were adapted to impose their own limit on the number and size of their breeding colonies (as an alternative to limiting the minimum size of individual breeding territories) they could combine optimum feeding conditions with maximum numbers.[24]

The Basel paper marked the beginning of Wynne-Edwards's commitment to the theory of group selection that he would present in 1962. In the intervening years he did not publish much. He was gathering his evidence and marshalling his forces. Although Elton's initial response to the Basel paper was subdued, he sent Wynne-Edwards an encouraging note in July 1955 after a visit to Aberdeen. After thanking him for his hospitality, Elton encouraged him to pursue his thesis, commenting, "We are all a bit tired of the present dogmas."[25] At this point Elton had spent more than twenty years working on the problem of animal population cycles and regulation. His Bureau of Animal Population had produced a dozen PhDs by 1955, and still the basic mechanisms and factors that regulate populations were ill-defined.

A Direct Confrontation of Ideas (Almost)

Wynne-Edwards's statement of theory, juxtaposed with the contemporaneous work of David Lack, led to a session at the centenary meeting of the British Ornithological Union (BOU) on population ecology. The conference was held at Oxford, and the proceedings were published as a special edition of the journal *Ibis* in 1959. The confrontation of the views of Lack and Wynne-Edwards was by proxy. Wynne-Edwards was in the United States as the Tom Wallace Professor of Conservation at the University of Louisville, Kentucky. The position was first offered to Julian Huxley, who turned it down. In October 1958 Max Nicholson, Wynne-Edwards's colleague from his Oxford days and from the Nature Conservancy, forwarded the invitation, which Wynne-Edwards ultimately accepted. From the outset, however, Wynne-Edwards was concerned about how his being in the United States would affect the possibility of his presenting a paper at the centenary meeting in March. He wrote to his friend William Thorpe, president of the BOU, asking for advice.

> As I think you know, I am working on what I may be forgiven for thinking is a tremendous theory, and have not far to go now before my book is finished. . . . If I do give a paper to the Centenary Conference I would like it to be on "The functions of social behavior," which would, I think, be revealed in a completely new light. It is always difficult to present a comprehensive topic in a few minutes, but I believe it would not be impossible to put across the general idea in the time allotted. You will see that this might be difficult for someone else to present in my stead, but I do not say it is impossible.[26]

Wynne-Edwards was also corresponding with William Clay, chair of the biology department at Louisville, angling for travel support to the conference. Although Clay felt the department could not fund the entire trip back for the meetings, he did offer partial funding and his encouragement. Discussions regarding Wynne-Edwards's attendance and participation at the centenary meetings also included David Lack, who along with R. E. Moreau was in charge of setting the scientific program. In a letter written on December 19, 1958, Lack mentioned that he had heard Wynne-Edwards might be in the United States for the meetings, and that if it was true perhaps there should be some rescheduling. Wynne-Edwards's reply acknowledged that he would indeed be in the United States at that time but that he planned to write his paper before he left and have somebody read it in his place.[27]

The person chosen to present Wynne-Edwards's paper was George Dunnett, a student of his who had also spent time at the Bureau of Animal Population at Oxford. He attended the meeting with David Jenkins, another former student of Elton's who was then working with Wynne-Edwards at Aberdeen. Given that the meeting was held at Oxford, the home-field advantage for Lack seemed to have played a role in the reception of Wynne-Edwards's paper. Both Dunnett and Jenkins wrote reports of the meetings to Wynne-Edwards, neither of them particularly encouraging. Before looking at these two reports, I will present a brief summary and analysis of both Henry Neville Southern's and Wynne-Edwards's papers as they were published in *Ibis*. The three other papers presented as part of the session on population ecology did not address population regulation and so are not relevant.

The paper written by Southern appeared first in the program and presented the two sides of the debate. Southern was a member of the staff at the Bureau of Animal Population at Oxford, which shared building space with the Edward Grey Institute for field ornithology, of which Lack was the director. His early work at the Bureau of Animal Population was concerned with the mouse as a wild animal. His most significant contribution to understanding the habits of the mouse was his discovery of its extremely small home range, and he was largely responsible for the third volume of *The Control of Rats and Mice,* the sourcebook of rodent control after the war.[28] Southern's later work, 1947 to 1959, focused on the predator-prey relationships of the tawny owl, and it was during this period that he published the article presenting Lack's and Wynne-Edwards's opposing positions regarding the causal factors in the regulation of animal populations. In "Mortality and Population Control,"[29] Southern presented two main points of view about the relative importance of birthrate and death rate in regulating bird populations. The first, that of Lack and others, held that birthrate was more or less fixed and that any adjustment was made by varying the death rate. The supporting argument is twofold. First, birthrate in birds is not flexible enough to raise or lower population level for adjustment to normally observed deviations. Second, clutch size is adjusted by natural selection to produce the maximum number of offspring parents can on average feed and send out in good condition. Therefore it cannot be significantly altered in immediate response to changes in population density.

The second point of view, supported by Wynne-Edwards, was that birthrate can be adjusted to population density, especially by declining when population numbers rise. Proponents point to three tendencies in bird behavior consistent with this point of view. One, many birds have small clutches. Two, it is common for birds to have long periods of immaturity. Three, many species of birds have long breeding seasons, sometimes over one year. All these tendencies, according to Wynne-Edwards, run contrary to the image of birds' breeding as rapidly as possible. According to Southern, supporters of this interpretation also argued that sociality and territoriality limit breeding and thereby reduce the birthrate.

Southern continued that it is artificial to separate birthrates and death rates in this way. From the point of view of population control, they are essentially one process that he suggested might usefully be termed repression.[30] Nevertheless, Southern was more sympathetic to Lack's view. He rejected Wynne-Edwards's argument using data from his own then current study of the tawny owl. Southern pointed out that the correlation between availability of prey and population size was sufficient to reject Wynne-Edwards's hypothesis.

In "The Control of Population Density through Social Behaviour: A

Hypothesis," read by Dunnett, Wynne-Edwards laid out the basic argument of his forthcoming book. It is also here that one can see hints of the kinds of analogies and metaphors he used that provoked such quick and largely negative response.

Wynne-Edwards asserted in the opening paragraphs that most animals, because they can move, play a predominant role in their own population densities. Although he acknowledged that food was almost always the critical limiting factor, he maintained that birds are largely successful in regulating their population densities below starvation levels through social conventions. It was also in this paper that Wynne-Edwards introduced the notion of epideictic behavior. The term applies to a very large class of social behaviors that appear to have evolved for the primary purpose of demonstrating population density. Their purpose, according to Wynne-Edwards, is to supply feedback into the homeostatic regulation of population density.[31] *The Oxford English Dictionary* notes only one use of *epideictic* before this article, specifically in the context of speech. Wynne-Edwards modified this usage to apply to the collective displays mentioned above. The *Oxford English Dictionary* cites three passages, two from Wynne-Edwards himself (1959, 1962) and the third from David Lack's 1966 book *Population Studies in Birds*. The juxtaposition of the quotations nicely encapsulates the authors' inimical positions, concluding with the following passage from Lack: "Most of the behaviour considered epideictic by Wynne-Edwards can be satisfactorily interpreted in other ways, and no positive evidence has been presented for an epideictic function."[32]

In the 1959 published version of his talk, the thrust of Wynne-Edwards's new hypothesis was illustrated by an analogy to human behavior. Citing man's own role as predator, he examined the problem of overfishing and argued that "under a system of free enterprise, he is in grave danger of impairing the resource by taking too big a harvest, depleting the stock, and entering upon a spiral of diminishing returns."[33] He went on to point out that the way to avoid this situation is through an agreement binding participants to limit the catch to the long-term optimum figure. This analogy is generalized as a natural and inherent attribute of the relationship between every kind of hunter and its staple prey. It is easy to understand how this analogy, which imputes very complex intentionality to members of bird populations, would arouse immediate suspicion on the part of naturalists working to further "scientize" natural history and ecological fieldwork.

In the rest of the same article, Wynne-Edwards sketched out how population density acts as a conventional buffer between the animals and starvation, that is, overexploitation of their food supply. In support of this assertion Wynne-Edwards argued that the territory system was capable of imposing a ceiling on population numbers. Furthermore, he argued, territories eliminate direct competition for food, and limited nest sites do not necessarily limit access to communal food sites but do restrict breeding. According to Wynne-Edwards, the territory system, in conjunction with social hierarchies or pecking orders, performs both of these functions.

Wynne-Edwards agreed with Southern's assertion that the separation of birthrates and death rates was artificial. He pointed out in his article that population density could be regulated through emigration of surplus members, but that this was most often achieved through modification of birth and survival rates of the population.

The final paragraph succinctly recapitulates Wynne-Edwards's argument:

> The hypothesis put forward here, therefore, suggests that animals have become adapted, with varying success, to control their own population densities, limiting them at the optimum level—this being the level that offers the best living to the largest number, consistent with safeguarding the food-supply from damage from so-called over fishing. It suggests that the result is achieved by interposing artificial, conventional goals as substitutes for direct competition for food.[34]

Given the preceding analysis from the published papers, it might strike one as surprising that when Dunnett wrote to Wynne-Edwards two weeks after the conference, he reported: "The population session, of which I'll give you a detailed account, was a flop. As was usually the case in the conference, there was very little time for discussion, which might have saved it, and what discussion did take place was, with the exception of one or two particular criticisms, vague, biassed and even irrelevant."[35]

Dunnett's letter detailed the day preceding the conference that David Jenkins and Adam Watson spent at Oxford. According to his account, both R. E. Moreau and Southern were "much opposed to your paper." At lunch another conference participant, James Fisher, told Dunnett "he was 'going to town' with criticisms of your ideas, saying (then) that if we were to believe this, then there was no longer any possibility in believing in Darwinism and natural selection!"[36] Dunnett went on to assert that "there is no doubt that these people, including I think, all the Oxford people, came prepared to object strongly to your paper."[37] He then speculates that

"Southern must surely have seen a draft in advance," adding parenthetically, "(I don't know this)."[38] This speculation is at least possibly true; David Lack had received a copy of Wynne-Edwards's paper fully two weeks before the conference, and as I pointed out earlier, the Edward Grey Institute, where Lack was director, and the Bureau of Animal Population, where Southern was a member of the small permanent staff, shared rather close quarters.

In his analysis of the papers presented in the session on population ecology, Dunnett concentrated most of his effort on the paper by Southern and on Lack's comments during the discussion.

> Mick [Southern] devoted the first half of his paper to an attack of your ideas, which by means of oratorical technique, seemed to carry the audience. It was not a nice or good attack, though one or two points he made were good. He criticised your ideas of a modifiable birth-rate and showed that it is generally rather inflexible (I think this is a valid criticism) and then referred to your earlier paper and said that low breeding rate and long immaturity were specific characters of the species concerned and were inflexible and not ecological adaptations. . . . Then he gave a great rigamarole about population "contouring" with respect to resource "contouring" as if this was a criticism of your ideas. I completely lost track here and so did David [Jenkins]. This section was confused, but, unfortunately, mainly by the use of slight sarcasm and his manner, he carried the audience.

Dunnett went on to mention James Fisher's criticisms, essentially dismissing them as insignificant. The tone of the letter indicates that Dunnett certainly felt that Wynne-Edwards's paper had not received a particularly fair hearing. Although the conference was attended by members of the BOU from all over the United Kingdom and beyond, the fact that the meeting was held at Oxford, among Lack's close colleagues and students, could not be ignored. As soon as Fisher sat down, Lack stood up,

> and without any preliminary gave a list of criticisms which were: 1) In his book most aspects of this were discussed at length ("dispersion and dispersal are Lack's terms") and book not mentioned. He had concluded that dispersion was not directly related to food. 2) Natural selection acts only on individuals. 3) Agreed with Kipling analogy. 4) Absence of facts: a) Territory: birds don't feed in them. b) Lack's Gt. Tit work gives opposite result from Kluijver and Tinbergen,

but not quoted; c) Colonial territory—no evidence; d) Peck order—Jim Lockie comes to similar conclusions in Lack's book. He then suggested more work on the study of food and population changes.[39]

When Lack sat down, discussion was closed and the session ended. All of the criticisms above are essentially straightforward and became the litany of objections that Lack would level against Wynne-Edwards for the next decade. The one objection that is not clear (number three) referred to a criticism raised by Fisher, where he likened Wynne-Edwards's idea of population control by territorial behavior to one of Kipling's essays on the Pribilof seal and dismissed it as romantic and improbable. It is not particularly surprising that Lack would concur that Fisher's analogy was apt. Immediately following the conference Lack sent Wynne-Edwards a short letter that included a copy of his comments at the conference. Lack wrote in his letter of April 30, 1959, "It was most kind of you to send me your paper, and I expect you will realise, I am in considerable disagreement with part, though not all, of it."[40] Unfortunately there is no reply to Lack's letter providing comments on his criticisms. This became something of standard procedure in the correspondence between these two scholars; the next letters do not refer at all to the disagreement between the two and concentrate solely on the examination of a doctoral candidate at Oxford. Dunnett closed his report on the conference with some thoughts about the future of Wynne-Edwards's theory.

> I can't help feeling that you are going to find it very difficult to put across these ideas to these people who are so obviously set against them. None of them seemed prepared even to consider the ideas involved but were much more concerned with poking holes in the examples given. For example it may not really matter whether territorial birds feed inside their territories or not, but the very fact that you implied that they do, is an opportunity for criticism on a factual basis. It seems to me that had there been no opportunity for criticising the examples, the ideas might have been considered, and a much better representation of your hypothesis would have resulted. I have tried hard to be fair in this, and have just read it on the phone to David, who agrees and has nothing to add except that he is sure that Mick rewrote his paper as a reply to yours.[41]

Given this account provided by Dunnett, the letter that David Jenkins sent to Wynne-Edwards a month later bears examination. Although he seems to be in general agreement with Dunnett's characterization of the events of the conference, his own recollection is not as damning of the Oxford response. He begins, "First, I don't think that there is any question of the Oxford people ganging up—they are too much divided among themselves for a start! E.g. Dennis Chitty's ideas and yours are remarkably similar. The impression I got was that you had put up some of these ideas before and Lack and Southern and [maybe] Moreau had made objections. They were now rather riled that you were putting up the ideas again without answering their earlier objections."[42] Jenkins was clearly less convinced than Dunnett was of the unified front of Oxford opposition. In his subsequent lines he concurs with Lack's objections regarding the synthesis of epideictic behavior and alterations in clutch size, and he also at least partially agrees with the criticism of some of Wynne-Edwards's examples. Jenkins concludes his analysis of the conference presentation and response with the following thoughts:

> Again I get the impression that because you introduce one or two questionable examples, e.g. that rookeries have feeding territories, the whole concept is rejected out of hand, indeed further is laughed to scorn. These are maybe strong words, but I think they are true. My own feeling is that your ideas on behaviour are absolutely right but as yet they are only ideas. The evidence for them is still rather weak and some of it is capable of alternative explanation. Of course I have not read your book but I fear that if the examples given are not water-tight, it may get a rather hard reception.[43]

The letters of Wynne-Edwards's surrogates at the centenary meeting were prescient. Although they differed in their sense of the organization of the Oxford opposition, they were in essential agreement regarding the degree of the opposition. The situation is similar to Darwin's immediately before the publication of the *Origin*. He and Wallace had jointly presented their theory to the Linnean Society, and Darwin was corresponding vigorously with Hooker, Lyell, and Huxley, gauging response and looking for converts and supporters. Furthermore, at that point it was not clear to what extent Darwin's closest colleagues were convinced of his theory. Wynne-Edwards's situation after the BOU meetings in 1959, one hundred years after the publication of *On the Origin of Species*, was strikingly similar.

Like Darwin, to have a chance of a fair hearing, Wynne-Edwards had to elaborate his ideas in book form and present them to a broad audience. In the next chapter I will examine the result of his commitment to group selection, his major *work Animal Dispersion in relation to Social Behavior.* This book became the touchstone for the debate between neo-Darwinists and group selection theorists for the next couple of decades.

CHAPTER FIVE

Animal Dispersion

Wynne-Edwards's Magnum Opus

By the early 1960s the modern evolutionary paradigm had coalesced to a significant degree. The major works of the modern synthesis had established natural selection as the primary mechanism of evolutionary changes and population genetics as the most useful method of analyzing and understanding these changes. In the centenary celebrations of the publication of the *Origin*, biologists and historians of science, among others, lauded the continuing robustness of Darwin's theory and its fundamental importance to every aspect of biological research.[1] As Theodosius Dobzhansky most famously put it, "Nothing in biology makes sense except in the light of evolution."[2]

Despite all the superficial harmony and official celebrations of the unity of biological research and theory, closer examination, not surprisingly, presents a more complex picture. In his 1992 history of ecology, Joel Hagen argues that it was at this time that the split between ecosystems ecologists and evolutionary ecologists dashed the dreams of a unified approach to ecology.[3] There was also the developing rift between the molecular approach, which was clearly gaining currency as an important new area of biological research, and the more traditional organismal approach.[4] While architects of the synthesis like Mayr and Dobzhansky were countering the vogue for molecular research, the last thing on their minds, particularly the mind of Ernst Mayr, an ornithologist and more traditional naturalist, was a new challenge from within his own area of research.[5] This indeed would be the challenge posed by Wynne-Edwards's forthcoming work.

Animal Dispersion in relation to Social Behavior, published early in 1962, was a work of epic proportions. It comprised twenty-three chapters, ran 653 pages, and covered a remarkable range of morphological and behavioral material throughout the entire animal kingdom. Wynne-Edwards essentially relied on the fieldwork of others for the many species he had not studied himself. He did, however, incorporate his own ornithological work and even his early marine studies into his grand theoretical framework. Contrary to the trend in biology at the time, his work did not include a great deal of statistical analysis or more sophisticated population modeling. Although Wynne-Edwards was clearly influenced and inspired by the work of Sewall Wright and Theodosius Dobzhansky, he did not incorporate their population analysis in any sophisticated way. Rather, he cited their work as evidence of a trend in evolutionary theory toward the study of population structure and distribution patterns that had been previously ignored. He wrote, "It has become increasingly clear in recent years, not only that animal (and plant) species tend to be grouped into more or less isolated populations, due very largely to the physical discontinuities of the habitat, but that this is a very important feature from an evolutionary standpoint in the pattern of their distribution (cf. Sewall Wright, 1938; Dobzhansky, 1941, p. 166 et seq.; Carter, 1951, p. 142)."[6]

Animal Dispersion did include a number of graphs and tables illustrating fluctuating population density and varied distribution patterns, but the theory and the evidence were largely presented as one long argument.

In the preface to *Animal Dispersion*, Wynne-Edwards claimed that for the past seven years, essentially since the presentation at Basel of the paper on low reproductive rates in seabirds, his theory of group selection had "provided [him] with a novel and, it has often seemed commanding viewpoint from which to survey the everyday events of animal behaviour; and some of the most familiar activities of animals, the purpose of which has never been properly understood, have readily been seen to have important and obvious functions."[7] As I argued in chapter 2, his realization of a theory to work by was consistent with some of his earliest work on behavior, including his work in the late 1920s on starling roosts, as well as the 1937 paper on the nonbreeding behavior of sexually mature fulmars. Wynne-Edwards took a second look at these obvious (in the case of roosting behavior) and puzzling (with regard to nonbreeding behavior) phe-

FIGURE 11. Gannetry at Cape St. Mary's, Newfoundland. Wynne-Edwards used this photo to demonstrate that populations demonstrated reproductive restraint. There is plenty of open space available, but the nests are mostly confined to the one guano-stained outcropping. Vero Wynne-Edwards Collection, Queen's University Archive, box 8. Photo courtesy of Queen's University Archive.

nomena and offered explanations that were significantly different from the prevailing explanation of the day. In 1929 he had admitted he did not have a conclusive answer for the roosting behavior of the starlings, but his suggested solutions were prescient.

In retrospect, the work with the fulmars had been even more interesting. This is a case where the phenomenon in question was just that. There was no clear agreement on whether these sexually mature adults were actually disengaged from the breeding cycle altogether or simply the result of inconclusive observation (that is to say, some of the individuals identified as nonbreeders may have been breeding individuals merely off the nest or otherwise engaged). Wynne-Edwards's study combined extensive observation to establish that these individuals were indeed nonbreeding with laboratory analysis of ovaries of nonbreeding females to establish that they were sexually mature and fertile. Having established both of these points, he then clearly identified their behavioral status as nonbreeders as a serious challenge to standard Darwinian interpretation.

This consistency of thought, from his early papers to *Animal Dispersion*, was commented on by more than one of Wynne-Edwards's colleagues, but perhaps most clearly in a letter from J. Z. Young. In a short note acknowledging receipt of his copy of *Animal Dispersion*, Young wrote: "I have just been reading your fascinating book. It has given me so many ideas that I cannot possibly tell you them all. I could not help thinking of some of your earliest studies on animal aggregations. It was interesting to me to see the continuity of your thought, even though you had perhaps forgotten it yourself."[8] As will become clear through this chapter and the next, this continuity of thought came to haunt Wynne-Edwards as his career progressed and his advocacy of the theory of group selection came to be seen as a personal crusade.

As might be expected, the initial reaction to *Animal Dispersion* in the correspondence was enthusiastic. Julian Huxley wrote, "I am just back from 5 weeks in the U.S.A., to find your great book waiting for me among a pile of other material. It is very good of you to send it to me. Last night I began reading it, and found it extremely interesting, not to say exciting; but I think I shall find myself in disagreement with some of your conclusions!"[9]

Charles Elton and Alexander Carr-Saunders also sent word of their appreciation for the book. Of course, Elton's professional opinion was spelled out more clearly in his review in *Nature*. A detailed analysis of this review will be presented later in this chapter. Before examining the professional critical response to *Animal Dispersion,* I will review its structure and content.

Wynne-Edwards, apparently somewhat consciously, modeled his work on what might be considered the archetype of the field, Darwin's *On the Origin of Species*. Writing in the preface of *Animal Dispersion* just over one hundred years later, Wynne-Edwards offers readers some insights into the development of the work as well as some warnings about a theory so wide-ranging and unconventional. He was clearly aware that it would stir up controversy, and the first paragraph ends with the statement, "Needless to say, the reader is confronted with two or three fundamental principles that, on account of their unfamiliarity alone, he may be expected to eye with a certain amount of scepticism, until they can by degrees be critically appraised in the light of each succeeding chapter."[10]

This book, it appears from these comments in the preface, was to be structured as "one long argument" linking the subjects of population and behavior. Wynne-Edwards described this structure as necessary so as to contain enough factual evidence to support the theory in its widespread ramifications.

Wynne-Edwards also noted that his new theory was fraught with philosophical implications. The examination of social behavior of animals, he claimed, provided the clearest indication yet of the "closeness of man's kinship with his fellow animals."[11] This invocation of the lessons about ourselves that we might learn from his new theory echoes not only Darwin's claim (as a result of his theory) that "light [would] be thrown on the origin of man and his history,"[12] but also James D. Watson and Francis Crick's concluding comment in their 1953 paper regarding the larger implications for the understanding of genetics. They wrote, "It has not escaped our notice that the specific pairing we have postulated immediately suggests a possible copying mechanism for the genetic material."[13] In each of the cases above, the authors have casually indicated their awareness of what they considered the major impact of their work. The understatements of both Darwin and Watson and Crick have become famous. Wynne-Edwards, on the other hand, who had hoped to rewrite the understanding of social behavior in terms of group selection, was never quite so successful.

By presenting Wynne-Edwards's prefatory comments here, I mean to suggest some similarity to Darwin's comments in the introduction to the *Origin*. One might argue that the commonality simply reflects a Whewellian approach, which calls for collecting myriad smaller facts in support of the larger truth; however, the reflection on Darwin does not end in the preface.

The first chapter of *Animal Dispersion* includes a recapitulation and expansion of Wynne-Edwards's argument by analogy with the overfishing example presented in the 1959 article. Wynne-Edwards suggested that the best approach to the subject of optimum density was to study man's own experience in exploiting natural resources. This analogy of human processes to natural processes clearly mimics Darwin's use of artificial selection as the analogue to natural selection in the *Origin*.

> The first chapter of Wynne-Edwards' book is the most interesting. It is here that he lays out his theory and presents its connection to previous thought about animal population and social behavior, as well as the fundamental differences he perceives to exist. In the second section of Chapter One, following the presentation of the over-fishing example, Wynne-Edwards asserted that something must constantly restrain populations, while in the midst of plenty, from over exploiting their resources. He rejected the application of such terms as "free enterprise" and "unchecked competition" to natural populations on observational grounds, and invoked instead the concept of the balance of nature.

> One of our first guiding principles, however, is that undisguised contest for food inevitably leads in the end to over-exploitation, so that a conventional goal for competition has to be evolved in its stead; and it is precisely in this—surprising though it might appear at first sight—that social organization and the primitive seeds of all social behavior have their origin. This is a discovery (if it can be so described) of the greatest importance to the theory.[14]

According to Wynne-Edwards a society is, in its most primitive function, merely an organization capable of providing conventional competition. This conventional competition precludes direct competition for food, or other resources, and thereby ensures the persistence of the social group by avoiding overexploitation.

Wynne-Edwards also took pains in the early pages of the book to differentiate his own theory from the traditional "Darwinian heritage" (read neo-Darwinism). He cited the standard interpretation of natural selection, which occurs at the individual (intraspecific) level and the species (interspecific) level, and argued that neither of these covered the social adaptations of interest to him. On his account, it takes a group of individuals to maintain social conventions. He cited the work of geneticists Theodosius Dobzhansky and Sewall Wright as supporting the notion that social grouping is of the utmost importance to evolution and the distribution of populations. Again, in chapter 1 Wynne-Edwards attempted to spell out the function of group selection.

> Evolution at this level can be ascribed, therefore, to what is here termed group-selection—still an intra-specific process, and, for everything concerning population dynamics, much more important than selection at the individual level. The latter is concerned with the physiology and attainments of the individual as such, the former with the viability and survival of the stock or race as a whole. Where the two conflict, as they do when the short-term advantage of the individual undermines the future safety of the race, group selection is bound to win, because the race will suffer and decline, and be supplanted by another in which antisocial advancement of the individual is more rigidly inhibited.[15]

One can see the parallels to the notion of the paradox of viability, introduced by Dobzhansky in his seminal *Genetics and the Origin of Species*, discussed here in chapter 4. Both Dobzhansky and Wynne-Edwards are essentially arguing that the mechanism of natural selection works on the composite of many individuals that form social groups, and that selection at this level is not merely the aggregate of individual selection; rather, it is an entirely different process whose results are unique and evolutionarily significant. It is important to point out here that for Wynne-Edwards essentially every population was a social group. An important part of the fundamental reassessment he advocated was a broadening of the understanding of the terms social behavior and social group. Previously, biologists had limited their use of these terms to a very definite list of species, especially the "social insects" and some higher animals. Wynne-Edwards pushed for an understanding of sociality that included nearly every species of the animal kingdom.

The last section of chapter 1 of *Animal Dispersion* describes the influence of Alexander Carr-Saunders. Unlike Darwin's Thomas Malthus, who was muse, or catalyst, or simply interesting reading during the early development of his theory, depending on whose account one chooses; Carr-Saunders's book, *The Population Problem*, was clearly an early influence on Wynne-Edwards. According to a recollection provided in chapter 1 of *Animal Dispersion*, at Charles Elton's suggestion Wynne-Edwards bought a copy of *The Population Problem* a few months before his final examinations at Oxford in 1927.[16] He rediscovered the book in 1959 after completing the first nineteen chapters of *Animal Dispersion,* leading to the satisfying discovery that much of his own theory had been anticipated by Carr-Saunders. For Wynne-Edwards, there were three most important aspects of Carr-Saunders's work. First was his claim that true nomadism did not exist in early humans; rather, tribes wandered within established territories. Second, Carr-Saunders argued that all human groups limited their fertility by a variety of practices including abstinence, abortion, and infanticide. Finally, he pointed out that every population has an optimum number, or optimum density, above which returns diminish. These aspects of Carr-Saunders's work on humans are completely consistent with Wynne-Edwards's own thinking about social groups, and he ended this section of chapter 1 by asserting, "Nothing could have given me greater reassurance than the knowledge that so distinguished a student had earlier pioneered the road."[17]

This connection to Carr-Saunders's work, and his ideas about group selection working on human cultural evolution and economics, was later developed by the Austrian economist and political philosopher Friedrich Hayek. In a recent paper, philosopher of science Erik Angner argued that

Hayek, who received the Nobel Memorial Prize in Economics in 1974, was more significantly dependent on the ideas of Carr-Saunders and Wynne-Edwards than on those of the moral philosophers traditionally identified as the source of his ideas on cultural evolution and economics.[18] Indeed, Angner's goal in the paper is to illuminate and clarify Hayek's reliance on evolutionary ideas and his debt to Darwin. This is particularly interesting given recent work in behavioral economics that once again invokes group selection models to explain various phenomena.[19]

The rest of *Animal Dispersion* is the catalog of facts promised in the preface. Again, like Darwin, Wynne-Edwards rarely cites others with regard to theory or interpretation, but he freely introduces their observations and fieldwork to support his own hypothesis. In fact, Darwin himself is cited sixteen times throughout the text, and most of the references are of this character. David Lack's work is cited more often than Darwin's (twenty-one times), and again the citations in general do not take issue with the theoretical aspects; rather, Wynne-Edwards reinterprets Lack's field observations and data to support his own theory. This was particularly provocative because of the stark differences in their interpretations of various behaviors. Wynne-Edwards acknowledged in the introduction to *Animal Dispersion* that "it was from Lack's stimulating book, and in particular the final chapter that the title of this one and the inspiration for writing it came."[20] It was not emulation that Lack inspired, however; the book was rather a direct challenge to Lack's interpretation.

After the introduction, Wynne-Edwards uses the next six chapters to examine the various contrivances animals use in transmitting and receiving social signals; there is a chapter dedicated to visual signals, a chapter for auditory signals, another for olfactory signals, and so on. Chapter 8 explores social hierarchies and the individual's role in them; here Wynne-Edwards is particularly dependent on the work of Warder Clyde Allee.[21] In chapter 8 Wynne-Edwards reemphasizes that all individual behavior is conditioned by the presence of other community members in the interest of regulating dispersion of the group as a whole. He also clearly redefines the theory of group selection: "Our general hypothesis that sociality is the basis of conventional behaviour, and that together they provide an indispensable part of the machinery required for the homeostatic control of dispersion, therefore merits very serious attention, because it identifies for the first time a possible common underlying purpose. We remember too how often already it has suggested a lucid meaning for what had previously been puzzling and unaccountable facts."[22]

Wynne-Edwards also goes to great lengths to establish that the social conventions fundamental to his theory are necessarily group characteristics, not apparent in individuals separately, but only when they form a group. He goes on to point out that in accordance with Lack, he recognizes food as the limiting resource, but the way this limiting resource affects a population is, on his account, completely different.

> It is part of our general theory that the free contest for food—the ultimate limiting resource—must in the long run end in over-exploitation and diminishing returns, and that this situation is avoided by substituting conventional rewards to take the place of actual food. Competition for these conventional rewards ideally operates in such a way that the population density is brought to an equilibrium at the optimum level—this being the level at which food resources are utilised to the fullest extent possible without depletion; and if the resource-level changes, or for any reason the density does depart from the optimum, forces are brought into play to restore the balance as early as possible.[23]

The difference between Lack and Wynne-Edwards is clearly presented. According to Lack's neo-Darwinian approach, food is the actual limiting resource, and the individuals that overbreed are less successful in fledging young. According to Wynne-Edwards's hypothesis, all this competition is conventionalized through the persistence of social groups.

> In practice, these conventions shield the resources against prodigal abuse from day to day by members of the contemporary population and thus safeguard the long-term prospects of group survival. It has already been pointed out in Chapter 1 that conventions of this kind must, by their nature, always be properties of a concerted group, and can never be completely vested in or discharged by a lone individual in perpetual isolation: their observance has to be reinforced by the recognition and support of others who are bound by the same convention. In the absence of parties they become meaningless.[24]

Chapter 11 provides a synthesis of communal nuptial displays, and arguments for group selection are presented in detail in chapter 12. Wynne-Edwards provides an interesting comparison between group selection and sexual selection in chapter 12, arguing that where adaptations evolve in relation to real situations, they are utilitarian and as economical as possible, but where the situations are conventional, the adaptations tend more toward the bizarre. This is the case both in Darwinian sexual-selected and

FIGURE 12. Epideictic behavior. Wynne-Edwards used these photos of a starling flight (*top*) and a swarm of whirligig beetles (*bottom*) as examples of behavior that signaled population density to group members. From *Animal Dispersion in relation to Social Behavior* (Edinburgh: Oliver and Boyd, 1962).

in group-selected adaptations. Wynne-Edwards uses the castes of social animals and stages in insects to support the theory of group selection in chapter 13 and chapter 14, and he presents many examples of communal roosts and hibernation consistent with his thesis. Chapters 15–20 describe a wide range of social behaviors covered by Wynne-Edwards's theory, including synchronization of social activities, twenty-four-hour and annual cycles of various animals, the vertical migration of plankton, interspecific associations, and mimicry; chapter 19 addresses the constancy of many of these epideictic phenomena, and chapter 20 deals with their variations. It is interesting to note some of the constancy of thought referred to in the letter from J. Z. Young cited earlier in this chapter. The transition between chapters 14 and 15 mimics exactly the subject matter of Wynne-Edwards's first two published papers, both of which appeared in 1929. The first, "The Behavior of Starlings in Winter," examined and attempted to explain the social roosting habit of this ubiquitous British bird; communal roosts and other social gatherings are the topic of chapter 14. The second paper Wynne-Edwards published in 1929 was a study in the *Journal of Experimental Biology* that attempted to determine the cues for the waking time of the nightjar; it is an early examination of periodicity, the subject of chapter 15 in *Animal Dispersion.*

The last chapters of *Animal Dispersion* (21–23) treat reproduction and Carr-Saunders's principle of the optimum number, mortality (especially juvenile mortality), and finally deferment of maturity and length of life span. Wynne-Edwards's habit of mining previous work is again evident in this final chapter. The plate from his 1937 article on the nonbreeding fulmars is reproduced to demonstrate the deferment of maturity.[25] Note also that the final section of the last chapter, on the life span, became the subject of an unpublished manuscript written three years before his death in 1997. "Senility as a Functional Adaptation" was never published but discussed the process of aging as a group-selected adaptation for maintaining population density.[26]

The Initial Reception

The publication of *Animal Dispersion in relation to Social Behavior* was a success for Wynne-Edwards. He was already a well-respected member of the community of animal behaviorists, and this book brought him world renown. By the end of 1962 he and his publisher had organized a trip around the world; Wynne-Edwards lectured in New Zealand and Australia, visited universities in Cairo and New Delhi, and watched birds and lectured in Bangkok and Kuala Lumpur. He and his wife also went to Fiji and Hawaii and visited former colleagues in Canada and Kentucky before returning to Aberdeen in mid-November 1962. Another indicator of the book's success was the offer from *Scientific American,* two years after the initial publication, to publish a précis, which sold 350,000 offprints.

Despite the book's popularity, the professional reception was decidedly mixed. In his short review in *Nature,* Charles Elton—a former mentor—took a rather dim view of the assurance with which Wynne-Edwards presented his thesis. In the opening of the review, Elton acknowledged that Wynne-Edwards had indeed identified a number of social behaviors that were not well understood but that by their very ubiquitousness called out for explanation. Elton also recognized the importance of developing a "satisfying theory about the natural regulation of numbers."[27] Elton, of course, had been working toward a theory of the regulation of animal numbers since the early 1920s and had created the Bureau of Animal Population at Oxford to meet these very ends, and so it might be expected that his response to so sweeping a work would be highly skeptical. Hence, after acknowledging the importance of the undertaking, Elton proceeded to criticize the book in rather strong terms:

> The theory is set forth with enthusiasm, often pontifically (if a bishop can wear blinkers), sometimes in a sort of messianic exaltation which admits of no other important processes affecting population levels. The language is usually lucid and the great array of facts that has been unearthed very interesting in itself; but the reasoning behind the language is extremely involved, and (in spite of the elaborate apparatus of decimal numbered paragraphs) rather woolly, and the new dogma incorporated in it safeguarded by many careful side-steps that make one rather uneasy, and certainly unconvinced at present.[28]

Elton's identification of Wynne-Edwards's thesis as "new dogma" is particularly interesting given his letter seven years earlier regarding the paper on low reproductive rates in birds that Wynne-Edwards had presented at Basel. In this letter (mentioned in chapter 4) Elton had expressed excitement about Wynne-Edwards's new line of thought and said that, as a group, population specialists had become "tired of the present dogmas."[29] This presents a dilemma for the research scientist: Should one risk obscurity by merely hewing to the contemporary norms or risk exile by challenging an established paradigm?[30]

In the rest of the review Elton is generous with praise for the amount of information Wynne-Edwards compiled into a single volume about the nature of social behavior and its coordination by means of behavioral signals, territorial patterns, and the methods by which animals determine and keep them. Nevertheless, he does not accept the major point of the book—that group-selected behaviors act as mechanisms to maintain populations below a threshold of overexploitation of food resources. Given that Elton's own work on populations was the result of numerous field and laboratory studies, his criticism of Wynne-Edwards's work as oversimplifying the way animals live was to be expected. Elton pointed out, "Most species fluctuate greatly in numbers, some of them violently, a few of them with remarkable regularity and it is difficult to imagine how there can be a set of behaviour signals elaborate enough to take care of all these different levels of numbers, or similarly of their fluctuating resources, so as to gear at all closely the population to their supplies of food."[31] But the disparity in their opinions about population fluctuations is not unusual. There were at the time fundamental debates about the nature of biological populations and whether they were generally considered stable.[32]

Finally, Elton asserted that the "enormous weakness of this enormous book" was that it provided no single case history of group selection. Although Elton allowed that group selection does occur and might be quite important, he was not at all convinced that it was necessarily related to population control. He concludes with the advice that the reader must decide for himself whether group selection occurs widely at all and that "once acclimatized to the persistent bias of the text, he will learn a great many unexpected things about animal dispersion and behaviour that certainly do require explaining as adaptations."[33]

Another review that was particularly critical of *Animal Dispersion* was written by the Danish ethologist F. W. Braestrup and published in *Oikos*. This review is of particular interest because the author states at the outset, "I am in perfect accordance with Wynne-Edwards." But this perfect accordance does not come without qualifications. Despite the perfect accordance Braestrup claims, the fifth paragraph begins:

> I think it may be said of the book, with a certain amount of truth, that most of what is sound is not new, and most of what is new is not sound. This in itself

> might not have been a very serious objection. If it had contained a good, well-digested and balanced summary of facts and views on animals' density regulation in relation to social behaviour, together with a clear statement of the problems (and on this foundation a statement of the more extreme theories of the author), it might have been a highly useful book.
>
> Instead, it is propagandistic with all the evil consequences of misplaced enthusiasm. It may of course be read in a leisurely way for the sake of the many interesting details given, but the reader who is eager to discover exactly what the author's theories are and what evidence he uses to base them on, will find himself frustrated and irritated long before he is through with it.[34]

Given the statement of accord that begins the review, these paragraphs are striking in their vehemence. This type of response signaled for Wynne-Edwards the possibility that he might spend the rest of his career as an outsider (or at least as long as he remained committed to group selection). Braestrup's review begins by putting Wynne-Edwards's recent work into context. He mentions his 1955 paper on low reproductive rates in birds, which he allows contained important arguments against Lack's view that in birds and mammals the number of eggs laid, or young born, corresponds to the largest number of offspring it is possible to raise. Braestrup concurs with Wynne-Edwards that Lack's view rests on a mistaken notion of natural selection. He writes that "the process of selection works, not only between individuals, but also between groups and between species, thus promoting characters which are to the benefit of the group, even in the face of contrary individual selection which may be kept in check by special devices. The existence of such evolutionary mechanisms is not only called for by a great many otherwise unaccountable facts, but is also in accordance with genetic evolutionary theory."[35] In support of his claim regarding genetics, Braestrup cites Sewall Wright's by now familiar review of George Gaylord Simpson's *Tempo and Mode in Evolution*, published in 1945.

Braestrup's first line of criticism comes in the form of a catalog of Scandinavian authors who preceded Wynne-Edwards in addressing the mechanism of group selection and its connection to social behavior. He cites some of his own work, as well as the work of O. Kalela and P. L. Errington. Although both Kalela and Errington are cited in *Animal Dispersion*, Braestrup suggests that these citations are insufficient and that readers who are truly interested in the subject matter would receive a much better primer from these authors than from Wynne-Edwards.

The second of Braestrup's criticisms is of a broader character, but it is particularly damning. Essentially, he argues that Wynne-Edwards mixes together various problems and concepts in an especially confusing way. For example, he cites Wynne-Edwards's use of the term epideictic display in lieu of more commonly used and noncommittal terms such as communal display. He argues that by characterizing various behaviors in this way Wynne-Edwards is perhaps suggesting to readers "that the highly controversial theories involved have actually been proved." Braestrup returns to this line of criticism later in the review when he considers Wynne-Edwards's inconsistent use of the term dispersion, which I will discuss below.

Another of Braestrup's general criticisms, which echoes Elton's, is that Wynne-Edwards has become overenthusiastic about his theory, giving him a peculiarly one-sided outlook. This enthusiasm has led him to push the beginning of competition for conventional goals very far back in the animal kingdom. While Braestrup is willing to acknowledge territoriality and social hierarchy in birds and mammals and even some insects, he is not convinced that it is applicable to barnacles and plankton, as Wynne-Edwards suggests. Interestingly, recent work on bacterial populations has led to the recognition of "quorum-sensing" behavior as a social trait whose function is to sense density and adapt accordingly.[36]

Finally, Braestrup's most serious criticism is reserved for Wynne-Edwards's assertion that this conventional husbanding of resources within groups could also obtain between species: "We now come, finally, to that part of the book, which, in my opinion, deserves the most violent opposition. I am alluding, of course, to the idea of an 'understanding' between competitors for the same resources. By inter-specific conventions an auto-regulatory mechanism for two or several species together should be built up, similar to that existing within single species."[37]

Here we see Braestrup essentially arguing that given Gause's hypothesis of competitive exclusion, which was well accepted at the time, the kind of interspecific arrangement Wynne-Edwards described is impossible. The first among the various problems Braestrup has with this idea is the question of what kind of evolutionary mechanism might bring about this mutual regulation. Second, he criticizes the inconsistent and ambiguous use of "dispersion." Having borrowed the term from Lack's *Natural Regulation of Animal Numbers*, where it meant, specifically, the nonrandom distribution of a species within a particular habitat, Wynne-Edwards is not so exacting. Moreover, according to Braestrup, Wynne-Edwards fails to distinguish between dispersion and regulation of numbers. Braestrup points out that "we thus have on the one hand dispersion *sensu strictu* [*sic*], and on the other hand dispersion in a wide sense, practically synonymous with regulation of numbers. On many occasions one has to simply guess which is meant, and the reader may be carried imperceptibly from one concept to another."[38] Finally, with regard to the notion of interspecific understanding hypothesized by Wynne-Edwards, Braestrup, after presenting various analyses of interspecific territoriality (especially in birds), asserts that "all known facts are against the existence or possibility of a 'syndication' of this kind—and the danger of over-specialization in any particular food is avoided in other ways."[39] This review was particularly upsetting to Wynne-Edwards, who kept several offprints in his personal collection.

The tone of these reviews was not typical, however. Most of the reviews were generally enthusiastic, and they offered varying degrees of support for Wynne-Edwards's application of group selection. Many of the reviewers congratulated him on the development of his thesis but warned against overextending its application. The review in *Ecology* by John King, professor of zoology at Michigan State University, is a good example. Near the end of his review King wrote: "The often proposed homeostatic function of epideictic displays in territorial birds has yet to be demonstrated beyond all doubt. That social behavior does affect dispersion, natality and mortality is recognized and Wynne-Edwards has done a service to bring the existing evidence together. However, by including practically all social behavior under the rubric of population regulatory mechanisms he has over extended his thesis."[40]

King wrote his review two years after the publication of *Animal Dispersion* and was thus in a position to comment on the book's reception by the professional community. King closed his review with the assessment that some had accepted the theory with an enthusiasm matched only by the author's, while others had rejected it out of hand. He ultimately suggested that some would use the theory to reexamine social phenomena and create hypotheses that would be more amenable to testing.

King's review was published the year before two major criticisms of Wynne-Edwards work came out. First, David Lack published *Population Studies of Birds,* which included three appendixes, the third one dedicated to refuting Wynne-Edwards. The second criticism was the now classic *Adaptation and Natural Selection*, by George C. Williams, which took as its target the form of evolutionary thinking that argued for selection above the level of the individual. These two works and Wynne-Edwards's response, as well as the increasing focus on the gene, are the topic of chapter 6.

CHAPTER SIX

Critique of Wynne-Edwards

Popular Reception of *Animal Dispersion*

The popularity of *Animal Dispersion* made Wynne-Edwards sought after as a speaker and raised his professional profile—an indication, perhaps, that some of the general ideas in his book translated well to a broader, nonscientific audience. The population bomb and environmentalism in general were major concerns of the 1960s, and Wynne-Edwards's theory of a naturally selected homeostatic population control was embraced by various political and environmental groups. From their perspective, he was making the very important point that all animals have a built-in social instinct and procedure, primarily geared to population control and the proper maintenance of natural resources.

In his review of *Animal Dispersion* in the *Nation,* David Cort supplies some rationale for the public interest. "It contradicts Darwin at some minor points, and is therefore in anguished controversy among biologists. It also constitutes by analogy a terrible warning to mankind, and that is why the press is toying with it so gingerly."[1] Cort continued in a more alarmist tone several paragraphs later:

> In Wynne-Edwards' proofs we can see reflected the breakdown of relations between parents and children, the male's and female's diminished attachment, the constant migration of peoples, the female's objection to being just a breeder,

> the male's resentment of being just a provider, smaller families, divorces, desertions, minorities escaping from "ghettos," elites struggling to keep out the invaders, the increase of homosexuality and neuroticism, alcohol and drugs, and above all, the evidence that young people, the group most sensitive to social stress, desire violence, especially if it is unprofitable and senseless: in all this we see that human society is reacting just as Wynne-Edwards says a crowded society should. It is giving a warning which nobody heeds; even when they see the Sunday cars jamming the highways, as in a dance of gnats, or a swarming of locusts.[2]

Cort's review echoes Wynne-Edwards's own overzealous application of the theory of group selection, in effect assigning "epideictic" function to each of these social ills. According to Cort's analysis, the breakdown of the family, confusion over social roles, and increases in crime are all evidence that humankind is carrying out some of the prescribed patterns of overcrowded animals. He goes so far as to say that this contribution of Wynne-Edwards "may have made all the social scientists superfluous."[3] It is not so difficult to imagine that given these kinds of claims and analyses, the attention to Wynne-Edwards's work, both positive and negative, would increase.

Before David Cort's review in the *Nation*, another review, on the other side of the Atlantic in *The Times Literary Supplement* (*TLS*), also heralded the importance of this new work. The anonymous reviewer begins by pointing out that the aim of pure science is to "fit our experience of the universe into a coherent pattern,"[4] and that the driving force of the scientist is the urge to make valid generalizations. With this in mind, he argues, Wynne-Edwards's book is to be considered of major importance, especially since the field of biology had very few generalizations to its credit at the time, with the notable exception of Darwin's theory of evolution by natural selection. Regarding *Animal Dispersion*, he writes, "Professor Wynne-Edwards has made a very broad generalization, and one that perhaps seems at first glance to be capable of meeting the essential test, of fitting together a wide range of facts, by no means all of them related."[5] Despite the breadth of the book, the reviewer writes that Wynne-Edwards makes no extravagant claims: "On the contrary, [*Animal Dispersion*] is so modestly written that the casual reader might almost overlook the fact that the author is producing a theory which, if true, offers an explanation of the origin and biological meaning of social life."[6] Wynne-Edwards's recognition of the near universality of "social" life is cited as one of the most important contributions of the work. The reviewer continues, "No longer, if he is right, can we think of [social life] as something that has cropped up unpredictably, and for no apparent reason in man, and here and there in other parts of the animal kingdom. We are bound rather to see social life whenever we look at animals, and we are presented, moreover, with a cogent reason for its universality. Social life in man, therefore, is no unique affair, but the culmination of a very widespread biological phenomenon."[7]

Following this introduction to *Animal Dispersion*, the reviewer's quotation can be recast as a question—whether Wynne-Edwards is right. From the reviewer's perspective, the reactions of ecologists and natural historians have been mixed. "Some, perhaps the more practically minded of them, have welcomed [Wynne-Edwards's theory] as a major advance. Others, perhaps the more theoretical or quantitatively minded, have been critical."[8] The reviewer, who appears more and more sympathetic to Wynne-Edwards's theory as the review progresses, argues that although many details of the book might be criticized and though some evidence can be better interpreted in other ways, the immediate question remains: Does the evidence add up to something that carries a measure of conviction? The reviewer answers yes.

The anonymous reviewer also points out that important to Wynne-Edwards's work is the realization that homeostasis is exhibited at the level of the group, or population. Homeostasis, the regulation of activity at an appropriate level, has been recognized as fundamental to the proper maintenance of cells, organs, and individual organisms, but it is now finally recognized at the group level: "It would be surprising if so widespread and indeed characteristic a feature of life did not also work to regulate the life of animals *en masse*."[9] David Lack takes up this point in his 1966 response to Wynne-Edwards, which I will discuss later in this chapter.

Toward the end of the review, there are three points of criticism. The first runs counter to the reviewer's earlier assertion that it does not really matter whether Wynne-Edwards's examples could be better explained by other theories. Simply put, Lack might be right, and food supply might be the actual limiting factor in population size. This is, quite obviously, a major criticism along the lines of "Wynne-Edwards could be utterly and completely wrong."

The second criticism is that Wynne-Edwards's analogy of natural predators to the commercial fisheries is seriously flawed and detracts from the strength of the argument rather than adding to it. The efficiency of a mechanized fishing fleet would allow it to deplete the resource much more quickly than animal predators, which might be expected to suffer from diminishing returns long before their prey was reduced to a level critical for its survival. This criticism was one of the first ever brought against Wynne-Edwards's theory, one that both David Jenkins and George Dunnett had pointed out after the British Ornithological Union conference in 1959.

The third criticism is especially interesting. The reviewer comments that Wynne-Edwards's theory "demands from the population or group of animals, a type of behaviour—in limiting reproduction—that runs counter to the general ideas of natural selection."[10] At first glance this seems less a criticism than an indictment of the whole project. Of course this theory runs counter to the standard neo-Darwinian account. That's the point. The reviewer, however, argues that although Wynne-Edwards was well aware that he was flouting scientific convention, he did not compensate with properly compelling arguments. The single, wider criticism of the book that the anonymous reviewer holds in reserve until the very end is that Wynne-Edwards's treatment of the subject matter is not sufficiently quantitative. He argues that "it was not until Darwin's theory was put on a mathematical footing by R. A. Fisher and others and subjected to rigorous experimentation by evolutionary geneticists, that it was properly established, and its implications fully understood."[11] The reviewer allows that this shortcoming is not Wynne-Edwards's fault; rather, it is the result of being in the vanguard of theory. Perhaps, he suggests, his most important contribution will be to lead to the development of the mathematical techniques necessary to assess his claims, and ultimately to "stimulate a more sophisticated approach to these problems."[12]

At this point it should be evident that Wynne-Edwards's theory had moved beyond the restricted community of evolutionary theorists and fomented intellectual interest in a broader context. It is also clear that, to this point, the reception of the theory was decidedly mixed.

David Lack and George C. Williams

The next two critiques I will discuss mark the shift toward a complete rejection of Wynne-Edwards's point of view and the solidification of the neo-Darwinian consensus. The first was by David Lack. In his 1966 *Population Studies of Birds*, Lack essentially recapitulated and then expanded much of the work of *The Natural Regulation of Animal Numbers*. The second yet more influential of the critiques was that of George C. Williams, presented in his classic *Adaptation and Natural Selection* (1966). The success of these two works is better understood in the broader context of the development of biology at the time.

The modern synthesis had connected Mendel and Darwin and reestablished the importance of natural selection as the primary mechanism of evolution. Simultaneously there was the contribution of the population geneticists, which emphasized the importance of population thinking. This emphasis was supported heartily by one of the most important evolutionary thinkers of the time, Ernst Mayr, who was also something of a mentor to David Lack. Nevertheless, Lack continued to pursue evolutionary theory from a strictly organismal perspective. Lack was following a crucial thread of the modern synthesis—organismal natural selection—but he ignored the theoretically important concept of population thinking. Although Wynne-Edwards had been criticized for the particulars of his presentation of the theory of group selection, clearly he was engaged in some important work regarding the concept of population thinking. Lack's work had skewed the lessons of the synthesis from neo-Darwinian (population-level explanations of evolution) back toward more Darwinian (organism-level) explanations. Lack then justified this move with an appeal to the principle of parsimony and claimed that his individual-level explanations treated the same phenomena that Wynne-Edwards's work attempted to explain at the group level and were therefore simpler and more correct. This simple argument is ultimately insufficient, as I will demonstrate in my analysis of Lack below. Regardless of the strength or weakness of Lack's parsimony argument, there is no question that the individual/organism-level explanations of Lack and Williams carried the day.

Lack's *Population Studies of Birds*

The idea of publishing a response to Wynne-Edwards occurred almost immediately after the release of *Animal Dispersion*. The renowned Cornell ornithologist Charles Sibley wrote to Lack on September 21, 1962. To a brief typewritten letter regarding the International Ornithological Union's upcoming congress, Sibley added a handwritten postscript: "Will you review Wynne-Edward's [*sic*] book? You should I think—it is going to be given favorable reviews by those who miss the point—and there will be many. Chris [Perrins] can tell you of our frustrating experience

FIGURE 13. David Lack and students at the Edward Grey Institute for Ornithology. This group hosted the 1963 meeting that discussed Wynne-Edwards's theory of group selection. Courtesy of Edward Grey Institute of Field Ornithology, University of Oxford.

with 'group selection' in Salt Lake at the AOU [American Ornithological Union] meeting. This matter needs to be exposed as the nonsense it is—and you're the one to do it!"[13]

In another letter written on February 15, 1963, just a few months later, Sibley expresses at some length his delight that Lack has decided to respond.

> I am pleased to see that you plan to answer Wynne-Edwards quickly. I truly believe that his book will be quietly dropped without causing much of a stir but nevertheless it would be well to have the nonsense clearly pointed out. I do hope you will include an emphatic statement relative to the fact that group selection is impossible on genetic grounds simply because only individuals, not populations possess genes. I think it is a mistake to enter into a long series of involved arguments about the interpretation of the facts. One should concentrate on this basic fallacy and force the group selectionists to recognize that they must invent a totally new type of genetic system before their arguments have any basis. They tend to gloss over this pitfall and tend to go blithely on saying, in effect, "Oh yes, but you don't quite understand" and so forth. They should not be permitted to leave this basic position until they have explained how the

> mechanism can possibly work with the type of genetic system evolved on this planet. All the rest is simply window dressing and nothing but a collection of interesting anecdotes misinterpreted on the basis of a false assumption right at the beginning. As always, I am fascinated to see how rapidly a person goes under and drowns as soon as he lets go of the firm rock of natural selection. In Wynne's case, the paradox is that he doesn't realize that he has let go of the rock but continues to believe that he has a firm grip even while he is resting under thirty fathoms of pointless examples.[14]

Sibley's encouragement and exasperation are illuminating. Clearly he does not agree at all with Wynne-Edwards's attempt to ally his own theory with the work of the population geneticists Dobzhansky and Wright. Furthermore, his concluding lines indicate that for him, natural selection can work only on the level of the individual organism. From his perspective, Wynne-Edwards has let go of the "firm rock of natural selection," but this is not so. According to Wynne-Edwards, of course, group selection is natural selection, consistent with the broader Darwinian intent but squeezed out by the neo-Darwinian focus on the individual. About the same time as Sibley's letter, Lack received another rather interesting opinion on whether he should respond to Wynne-Edwards's group selection theory and, if so, how. This one came from Robert MacArthur, an American ecologist and theoretician with a much more mathematical approach than Lack's, who had spent his first postdoctoral year at the Edward Grey Institute to receive additional training in field ornithology. Lack was an important influence on MacArthur, and his letter shows deference to a mentor while cautioning that his conclusions regarding group selection might be premature. In a letter from early October 1962, MacArthur was careful in his advice. After a brief paragraph regarding his subscription to the *Journal of Animal Ecology,* he revealed his primary reason for writing. He had heard from Evelyn Hutchinson and Gordon Orians that Lack was planning a book that would, at least in part, refute Wynne-Edwards's group selection theory. He wrote, "May I take the liberty to urge you not to." He continued,

> My reasons are two: First, I think the quickest and surest fate of incorrect science is oblivion. You remember Huxley's advice to Darwin over the Butler affair. Second, and more important, I am sure that an analysis of group selection would be premature. Although I am pretty much on your side in this affair, I can

> see only two possible ways of disproving group selection. Either we show there is no possible mechanism or we must show that nature never uses the possible mechanism. The first of these is definitely wrong; there are possible mechanisms. The only doubtful point is how large a role they play in nature. And that analysis must wait until someone lists some phenotypic characteristics on which individual and group selection would act in unambiguous and different directions. . . . I feel a little silly writing this because I may have misunderstood your aims completely. Please forgive me if I have.[15]

Lack responded that MacArthur had not misunderstood his intentions and pressed him with some technical questions on the possible mechanisms for maintaining an altruistic gene in a population by group selection. Ultimately, consistent with Sibley, Gordon Orians, John Emlen, and other supporters, Lack decided to go forward with the response to Wynne-Edwards and to challenge group selection.

In his critique of Wynne-Edwards, which appeared as an appendix in *Population Studies of Birds*, Lack cited the *TLS* review and added a question of his own: Why, if all the criticisms he had provided were correct, had Wynne-Edwards's book garnered any support at all? Lack provides three basic reasons. First, some of Wynne-Edwards's views had been current for some time in the evolutionary and ecological literature, but more implicitly than the way they were presented in *Animal Dispersion*. An explicit treatment of sociality, matched by an unprecedented comprehensiveness, constituted the most obvious reason for the book's popularity. The second and third reasons were cited by the anonymous reviewer:

> Any biologist is bound to be attracted by two very general aspects of this thesis: first, . . . the universality and the underlying reason for social life; and second, that the theory finds a place in animal populations for the phenomenon that biologists call homeostasis—the regulation of activity at an appropriate level. Homeostatic mechanisms . . . have long been familiar to physiologists. In recent years comparable devices have been found to reign within individual cells, and in the innermost working mechanisms of heredity. It would be surprising if so widespread and indeed characteristic a feature of life did not work to regulate the life of animals *en masse*.[16]

In the final two paragraphs, Lack dismissed Wynne-Edwards's work in a couple of deft strokes. Essentially, he argued that the positive response general readers had to Wynne-Edwards's book could be explained by a combination of ignorance on the part of the audience and obfuscation on the part of the author.

> Both these reasons rest on ignorance by other biologists of what ecologists have already discovered. First, various ornithologists have long been aware of many displays and other social behaviours in birds which are not to be explained in terms of sexual selection; but rightly, in my view, they have ascribed various functions to them and not a single overriding one (epideictic). However, the general reader might not be aware of this from Wynne-Edwards' book, as he did not usually discuss the earlier interpretations of the phenomena which he considered epideictic. Secondly, it is of course true that natural populations could not be regulated without homeostasis, and though a few workers like Andrewartha and Birch have claimed that natural populations are not regulated, most modern workers agree with A. J. Nicholson that they are, primarily through density-dependent mortality factors, though dispersive behaviour plays an important secondary role. Ecologists have not used the term homeostasis, but the idea of self-balancing populations has been widely accepted by them at least since the paper by A. J. Nicholson in 1933.[17]

Lack concluded that there is indeed a phenomenon of dispersion that is fundamental to maintaining the proper population density with respect to the food supply, but that it is not as extensive as Wynne-Edwards claimed. Furthermore, most of the behavior Wynne-Edwards considered epideictic could be satisfactorily and more parsimoniously explained in other ways. This shortcoming was compounded, according to Lack, by the persistent lack of supporting evidence in *Animal Dispersion*. Lack also argued that there is no reason to suppose that reproductive rates have evolved to balance mortality rates; again, reproductive rates are sufficiently explained by orthodox natural selection and the balance between birthrates and death rates that Lack attributed to density-dependent mortality. Finally, low reproductive rates, one of Wynne-Edwards's earliest cases for group selection, can, on Lack's account, be fully explained by natural selection without recourse to group selection.

In a notebook Wynne-Edwards began in 1971, in which he kept his notes on articles regarding group selection in preparation for his second book, he made a page-by-page notation of the criticisms contained in appendix 3 of Lack's *Population Studies of Birds*. The argument in the appendix became a standard source for the rejection of Wynne-Edwards's work, but as my analysis will show, the shortcomings of the arguments in the appendix are even more apparent in the body of *Population Studies*. Unfortunately, the lists Wynne-Edwards included in his notebooks cataloged points to be addressed rather than providing substantial responses to the criticisms listed. On page 29 of "Black Book Number Four" Wynne-Edwards wrote:

> Pp. 300–302. I must show that evidence for the absence of "overfishing" in nature is not illusory, as stated on p. 302.
>
> p. 303. He says no need to regard the hierarchy as a group character (see p. 284 of this same book for summarized reasons, which are, in short that it is of ultimate benefit to a subordinate not to challenge or invite attack from stronger individ. but to seek food somewhere else)
>
> p. 305 Points out a weakness of Jespersen's correlation as explained by me.
>
> p. 308 He says "man is so influenced by tradition that any parallel with the social behaviour of animals is highly dangerous." (At foot, misrepresents what I said about correlation / longevity & reprod. rate.)
>
> p. 310 He says "I consider that the primary factor influencing adult numbers is density-dep. mortality outside the breeding season." I must point out that mortality outside the breeding season is secondary; primary cause is social exclusion.
>
> p. 311 Uses "homeostasis" to describe a passive effect or consequence of extrinsic stability forces & not an active intrinsic process.[18]

Reading through this abbreviated list, one wonders if this is all the response Wynne-Edwards could muster to Lack's criticism, since at least five years had elapsed between the publication of *Population Studies* and the creation of these notes. Indeed, Wynne-Edwards's response to his critics, or lack thereof, was to become a critical aspect in the reception of group selection theory over the next two decades, and it was of particular importance in the response to Wynne-Edwards's second major book, *Evolution through Group Selection.*

In fact, the charge that Wynne-Edwards failed to respond to his critics was preceded by a closely related criticism that forms the basis of most of David Lack's response to *Animal Dispersion*—that Wynne-Edwards failed to acknowledge his predecessors' explanations of the same or similar phenomena, and that he offered too little evidence to support his alternative interpretation. Although the appendix is Lack's most tightly argued response to Wynne-Edwards, his analysis and appraisal begins on page 2 of *Population Studies* and continues throughout the text. The first critical mention of Wynne-Edwards comes on page 7, and this passage illustrates the nature of the entire dispute between the two. The subject here is reproductive rate, and Lack points out that in his 1954 book he provided circumstantial evidence that the number of broods per year is as large as the environment permits. He also argued that each species begins breeding when it can do so without undue risk to itself. Wynne-Edwards, by contrast, argued that deferred maturity had evolved in long-lived species via group selection as a mechanism for avoiding overpopulation. Wynne-Edwards also held that the reproductive rate in most species had evolved, again through group selection, to balance their mortality rate. Lack's response is typically swift and dismissive; however, he equivocates a bit near the end of his condemnation of Wynne-Edwards's position:

> But the balance [between reproductive and mortality rates] is as readily explained through density dependent variations in the mortality rate, and there are theoretical reasons for thinking that certain mortality factors, notably food shortage, predation, and disease act in a density dependent manner. When I wrote my earlier book of 1954, the existence of density dependent mortality still rested largely on theoretical considerations, supplemented by data from laboratory populations of various insects, which were, however, models rather than true experiments. *The evidence from natural populations is not much better now, but nevertheless I believe that density dependent mortality provides the best explanation of the balance between birth and death rates.*[19]

This is an interesting passage, especially the concluding lines, because it lays bare Lack's commitment to a particular theoretical construct despite a persistent lack of deciding evidence. Given that Lack and others had been working on these questions at least since the publication of *The Natural Regulation of Animal Numbers* in 1954, one might reasonably expect an increase in the data supporting the theory. By the same token, an individual who was not sympathetic to that particular paradigm might seem justified in offering an alternative interpretation of matters already "explained." This is the position of Wynne-Edwards while he was working on *Animal Dispersion*. What is perhaps more interesting is that this position is apparently justified, and this justification, ironically, is provided by Lack's book. The reiterated criticism in *Population Studies* is that Wynne-Edwards is dealing with behaviors that have already been accounted for, and that his new explanations are not supported by sufficient evidence. In nearly every one of the passages directly critical of Wynne-Edwards, Lack writes that Wynne-Edwards's interpretation is unnecessary or incorrect and follows that claim with the admission that there are too few data to reach a firm conclusion. Despite the lack of evidence, the received view is that Lack's criticisms of Wynne-Edwards theory were devastating and that Lack carried the day; here again a broader context may help to explain the state of affairs.

The differences between Lack and Wynne-Edwards are stark. Lack's experience emphasized morphology and biogeography, consistent with the interpretations of Ernst Mayr and others. Wynne-Edwards had been trained by Elton and had always been driven by questions about populations. Lack's individual-level explanations struck a chord with the professional biological community, which continued to look for straightforward, lawlike application of evolutionary theory. His work provided exactly these kinds of explanations.

Another of the weapons in Lack's critical arsenal was Occam's razor—the assertion that an existing explanation should not be replaced by one that is more complex. He used this approach in a section on the European blackbird.

> The seasonal variations in clutch-size must be allowed for when examining the possible influence of other factors. When this is done, it is seen that in the Blackbird, as in the Great Tit, the average clutch is larger in the woods than gardens and larger in older hens than yearling hens. The reasons for the difference in habitat require more study and might, partly at least, be linked with the fact that the population density is much lower in woods than gardens. The difference due to age is, I suggest, adaptive and due, as in the Great Tit, to older parents being able, on the average, to raise rather larger broods than those breeding for the first time. On the other hand Wynne-Edwards attributed it to his idea that fertility is governed by the interplay of social rank and economic conditions, with the result that newcomers to the breeding caste are liable to be handicapped by their inferior social position. He provided no evidence for this, nor is any to be found in the studies of Snow on the Blackbird or Perrins on the Great Tit, but both these latter authors had evidence that older parents could raise more young than one year old parents fitting my simpler explanation.[20]

Several interesting things are going on in this passage. Lack admits to a softness of the data but asserts that his position is correct. He also appears to conflate the simplicity of an explanation with its familiarity. Finally, it is important to note that both Perrins and Snow are products of Oxford and members of the Edward Grey Institute, so it is not surprising that their work might support the position Lack advocates here. Returning to the first point mentioned above, one can see how Wynne-Edwards might use the same data Lack provided to support his own position. That the older pairs are more successful in raising young is completely consistent with either account. In fact, Wynne-Edwards cited Lack on this point in *Animal Dispersion*.

> An alternative explanation has been suggested by Lack, namely that breeding imposes a strain on the parents, which may prove too heavy for the younger individuals in these slowly maturing species, so that they tend to succumb if they attempt too soon; natural selection has consequently retarded the onset of maturity, giving them time to develop the stamina required to cope with it. *This theory does not enable us to account for the very wide spread in the age of first breeding that we have seen to occur in individuals of the same species*; presumably the same kind of selection would work about as strongly against individuals that were handicapped by coming to maturity a year or two behind the average, and would result in fixing the age of puberty within rather narrow limits. *Nor does it account in any way for the fact that in so many cases there is a reserve of sexually potent, but actually non-breeding, individuals, ready to step in and take the place of any breeding bird that is killed.* As Lack was aware, *it fails again to suggest why the males should often be retarded longer than the females, when if anything, the former have the lighter reproductive task.* Finally, there is no evidence to support the basic proposition that mortality is higher among breeders than among non-breeders in the younger year classes.[21]

In the three passages I have emphasized in the quotation above, Wynne-Edwards raises some challenges to Lack's interpretation that are not answered in *Population Studies*. First, he argues that the theory does not explain the wide age range for first breeding. If indeed this trait is fixed by natural selection acting at the individual level, one would fairly assume that this age would be more precisely fixed. Second, Lack's account fails to account for the presence of nonbreeders, which had attracted Wynne-Edwards's attention in 1937. Finally, Wynne-Edwards complains that Lack failed to explain why sexual maturity would be deferred in the males more often than in the females, when the males carry less of the reproductive burden. This kind of disagreement is typical of the disputed points that make up the bulk of references to Wynne-Edwards's theory throughout Lack's book. Unfortunately, it also reflects the kind of talking past one another that would characterize the group selection debate for the next several decades.

Deferred maturity was perhaps the most contentious of the issues between these two naturalists, and Lack uses the example of the white stork to counter Wynne-Edwards's argument; notice that again no determining evidence is provided.

> My suggestion that deferred maturity has been evolved in those species of birds in which the raising of young is so difficult that it would impose undue strain on inexperienced younger individuals has been rejected by Wynne-Edwards, particularly on the grounds that, in the White Stork and other species, the age of first breeding varies. On my view, he claimed, natural selection should have resulted in a more or less fixed age of maturity. But in my view variation could be expected either if the age of first breeding can be phenotypically modified to the food situation, as suggested above, or if the advantage of starting at one particular age or a year younger were on balance almost equal. Wynne-Edwards produced no evidence for his alternative view that "the homeostatic machinery allows only the appropriate number of individuals to breed in any given area in any one year," thus preventing overpopulation, and there is nothing in the published work of Shuz or others to suggest that this is what happens.[22]

This passage again demonstrates clearly that the evidence for either of the two interpretations might have been perceived as legitimate. Nevertheless, Lack concludes his section on the white stork with the suggestion "that the concept of prudential restraints in birds may be rejected, and it is pleasing that the data for rejecting this idea come from so well-known a fertility symbol as the White Stork."[23]

In a later chapter on mutton birds and the Manx shearwater, Lack also rejected Wynne-Edwards's explanation of slow growth rates and long incubation periods as group-selected adaptations for maintaining population levels, and he again attributed it to "sparse and variable food supply." He admitted in the following sentence that he was "not very happy about this explanation but can think of no better one."[24] Lack then went on to argue that Wynne-Edwards did not acknowledge this explanation either in his 1955 paper on low reproductive rates in seabirds or in *Animal Dispersion*. But later in the same chapter Lack argued against the behavior that got Wynne-Edwards started on the path toward group selection—the intermittent breeding of the fulmar. Lack asserted, with regard to intermittent breeding, that "[Wynne-Edwards's] evidence was indirect and is not, in my view, convincing, nor did it hold for a population of marked individuals of the same species in Scotland."[25] Lack, however, did not cite the original paper on this very topic that Wynne-Edwards published in 1937. (See my discussion of this paper in chapter 3.) In the penultimate chapter of *Population Studies*, he was again forced to admit that there is no satisfactory determining evidence either for his position on deferred maturity in long-lived species or for Wynne-Edwards's position. Toward the end of this chapter Lack argued,

> It should be added that deferred maturity will be of particular value in long-lived species which lay only a single egg, like shearwaters and albatrosses, as the omission of a year of potential breeding will make little difference to the eventual number of offspring produced. Contrary to my view, Wynne-Edwards considered that deferred maturity is yet another means of bringing the reproductive rate low enough to prevent over-population in such long lived birds. *There is not as yet enough evidence in Procellariiformes to discuss his or my view critically.*[26]

The concluding line here is particularly interesting given that Lack, in the preceding pages, did indeed discuss Wynne-Edwards's view critically and essentially rejected it.

A final example should help make it clear that despite the certainty and repetition of Lack's criticisms, he himself allowed that the evidence in most of the examples concerned was not conclusive. In the final chapter of *Population Studies* Lack makes another pass at the topic of deferred maturity. Here, most surprisingly, where Lack might be expected to try to put the issue to rest, he made the most qualified statement about the support for the alternative interpretations.

> There is no evidence for the view of Wynne-Edwards that such deferred maturity has been evolved through group selection in long-lived species to reduce the number of young and so prevent over-population. There is equally no proof of my alternative view that, in such species, breeding is difficult and individuals which try to breed when younger than the normal age leave, on the average, fewer not more surviving young than those which start later.[27]

These statements indeed reflect a situation that was unresolved. They might have been seen, especially by Wynne-Edwards, as a potential paradigm shift with regard to animal sociality. Of course, this was not the case. Lack's 1966 book was not received with the enthusiasm of the 1954 work, but that may have been because much of it was similar to the positions taken in *The Regulation of Animal Numbers*. I think that, after examining Lack's criticism here in some detail, one might have a better understanding of Wynne-Edwards's persistence and of why he continued to struggle against the increasing tide of criticism of his theory.

The final paragraph of Lack's book offers a prescription for future researchers based on a presidential address Thomas Park gave to the Ecological Society of America in 1961. Lack quotes Park as stressing that "natural history . . . is one of the prime sources of insight and knowledge for the modern ecologist. It helps him visualize a problem and ask a cogent question." Second, Park continued that while "traditionally ecological findings are based on the observation of events which have taken place in an environment unmolested by an observer and varying to its own natural right . . . I am persuaded that our research progress, and the validity of our interpretations, will be enhanced whenever we intelligently modify appropriate elements of the environment by predetermined plan."[28]

Lack's citation of these lines from Park is interesting for two reasons. First and foremost is Park's affiliation with the Chicago school of ecology, which emphasized the study of the population as a distinct physiological and evolutionary unit, and its concurrence with the idea of group selection and in particular the notion of the physiology of populations. There could not have been a group of ecologists less sympathetic to the orthodox neo-Darwinism expressed by Lack than the core of the Chicago group, Warder Clyde Allee, Alfred Emerson, and Thomas Park. The Chicago school was incredibly influential in the developing field of ecology. Its emphasis on the role of the group and the concept of the superorganism was not readily accepted by the broader biological community. In his 1992 book *The State of Nature*, historian Gregg Mitman provides an excellent analysis of this group of researchers.[29] Thus it is indeed ironic that Lack should cite Park to inspire future ecological work.

The second reason the positive citation of Park is noteworthy is that George C. Williams, in the new preface to *Adaptation and Natural Selection,* cited a lecture by Alfred Emerson, Park's colleague at Chicago, as the catalyst for his effort to banish this type of thinking from the evolutionary lexicon:

> My first awareness of a motive for *Adaptation and Natural Selection* came during the 1954–55 academic year while on a teaching fellowship at the University of Chicago. The triggering event may have been a lecture by A. E. Emerson, a renowned ecologist and termite specialist. The lecture dealt with what Emerson termed beneficial death, an idea that included August Weismann's theory that senescence was evolved to cull the old and impaired from populations so that fitter youthful individuals could take their places. My reaction was that if Emerson's presentation was acceptable biology, I would prefer another calling.[30]

Williams continued to argue that a broader dissatisfaction arose from the overall inconsistency in the use of the theory of natural selection, and his book would be an attempt to set this right. On the other side, Williams cited Lack as a guide to the proper application of the theory of natural selection. In particular, Williams claimed that Lack's 1954 article on the evolution of reproductive rates was "a sublime encouragement. I had found a biologist who believed, as decisively as I did, that natural selection is a real scientific theory."[31] The article Williams cited was published in the collection *Evolution as a Process,* edited by Julian Huxley, A. C. Hardy, and E. B. Ford. It contained a précis of the argument regarding reproductive rates that Lack was to publish that same year in *The Natural Regulation of Animal Numbers.*

Williams's *Adaptation and Natural Selection*

The introduction to Williams's book sets up the past one hundred years of biology as a contest of two groups: those who emphasize the role of natural selection as a primary or exclusive creative force, opposed by those who minimize its role in relation to other proposed factors. According to Williams, this contest was decisively won in 1932 when R. A. Fisher, J. B. S. Haldane, and Sewall Wright published their major works. Williams's concern was that contemporary evolutionary theory derived from the same sources that led to the discredited theories of the nineteenth century. The major problem facing evolutionary theorists was the correct identification of adaptations.

To address this problem, Williams set out to provide a set of ground rules and demonstrate how to use them. The first rule was to recognize adaptation as a special and onerous concept that should be used only when necessary. Further, when an adaptation was recognized, it "should be attributed to no higher a level of organization than is demanded by the evidence."[32] In the second chapter Williams takes as his first target the notion of evolutionary plasticity. As I pointed out in chapter 3, this idea was fundamental to Dobzhansky's ideas about evolution and provided some inspiration to Wynne-Edwards in his early thinking about population-level traits. Williams essentially argued that natural selection would always act to maximize the mean reproductive performance, regardless of the effect on long-term population survival.

In chapter 3 of *Adaptation and Natural Selection*, Williams discussed the various environments that play a role in the relationship of genotype to phenotype. Williams allowed that multiple environments played essential roles in determining selection coefficients. Nevertheless, he stressed that "the recognition of this fact in no way compromises the principle of selective gene substitution as the sole and ultimate source of adaptive evolution." He went on to argue that natural selection will always favor rapid development, counter to the work of Wynne-Edwards and others and consistent with the arguments presented earlier in this chapter by David Lack.

Chapter 4 is perhaps the most relevant here, as it is directly concerned with group selection. Williams's first order of business was to set out group selection as necessarily working opposite to genic selection. Although Wynne-Edwards discussed many such cases, he did not argue that these two evolutionary forces were always at odds. Second, Williams addressed the difficulty of assessing a population's success. Individual fitness could be measured in rather straightforward ways; populations, on the other hand, required a much more complex and less satisfactory method. Despite this difficulty, Williams settled on numerical stability as the best indicator of population success, consistent with Wynne-Edwards. He did not, however, accept that this stability results from "biotic adaptations." The challenge for the biologist here is to determine if the survival of the population is the result of the adaptedness of the population or the adaptedness of the individuals that constitute it.

> We must decide: Do these processes show an effective design for maximizing the number, rate of growth, or numerical stability of the population or larger system? Any feature of the system that promotes group survival and cannot be explained as an organic adaptation can be called a biotic adaptation. If the population has such adaptations it can be called an adapted population. If it

> does not, if its continued survival is merely incidental to the operation of organic adaptations, it is merely a population of adapted insects.[33]

This distinction between organic and biotic adaptations is key for Williams. Introduced at the beginning of chapter 4, this concept is useful for avoiding the semantic difficulties that accompany many of the debates regarding natural selection. On this account, an organic adaptation is a mechanism designed to promote the success of an individual organism as measured by extent to which it contributes genes to later generations of the population of which it is a member. A biotic adaptation is a mechanism designed to promote the success of a biota, as measured by the lapse of time to extinction. For Williams, one cannot have group selection without biotic adaptations, and the rest of the book is dedicated to rejecting their existence.

In chapter 5, "Adaptations of the Genetic System," Williams pointed out that biologists commonly regard sexual reproduction as a biotic adaptation. One of the functions attached to this adaptation was maintaining evolutionary plasticity, Williams's first target in his critique. He rejected this function for group selection on the grounds that there were perfectly adequate explanations at the individual level, precluding the need for higher-level explanations. Again in chapter 6, Williams pointed out that much of the work regarding higher-level selection was the result of sloppy thinking. In this case he pointed to the distinction between results and functions. Species survival, for example, is one result of reproduction. But this is not evidence that species survival is a function of reproduction. "If reproduction is entirely explainable on the basis of adaptation for individual genetic survival, species survival would have to be considered merely an incidental effect."[34] According to Williams, the group selectionists' insistence that species survival is the result of selection acting on groups is founded on the misconception that individual reproductive success would be maximized by unbridled fecundity. Here again Williams cited Lack's work that corrects this erroneous view.

In the next chapter, on social adaptations, Williams argued that most of the animal behavior patterns between unrelated individuals are manifestly competitive, not withstanding the work on cooperation by animal ecologists. He also argued, based on his own work and that of William D. Hamilton, that the absence of the comparable organization (comparable to that of the social insects) in any group of unrelated individuals is cogent evidence of the unimportance of biotic adaptation (read group selection). Finally, Williams criticized much of the thinking with regard to group selection for relying on aesthetics and attempting to derive morality from nature. Again, this point of view was particularly true for the group of Chicago ecologists that Williams had worked with in the late 1950s. From Williams's point of view, the misperception that the population was an evolutionary unit had to be eradicated. According to his reasoning, the population should be regarded as part of the environment that the individual is adapted to, not something that is an adapted unit.

A return to the preface of *Adaptation and Natural Selection* is useful here, given all that Williams has argued regarding group selection. In the 1996 preface he wrote:

> A few years after 1966, I was being given credit for showing that the adaptation concept was not usually applicable at the population or higher levels, and that Wynne-Edwards' thesis that group selection regularly leads to regulation of population density by individual restraints on reproduction was without merit. It also became fashionable to cite my work (sometimes, I suspect, by people who had not read it) as showing that effective selection above the group level can be ruled out. My recollection, and my current interpretation of the text, especially Chapter 4, indicate that this is a misreading. I concluded merely that group selection was not strong enough to produce what I termed biotic adaptation. . . . [This] would be characterized by organisms playing roles that would subordinate their individual interests for some higher value, as in the often proposed benefit to the species. Even without its producing biotic adaptation, group selection can still have an important role in the evolution of the Earth's biota.[35]

This clarification regarding his position on group selection has been recently emphasized by Elliott Sober and David Sloan Wilson in *Unto Others*. In fact, both Williams and Hamilton, whose contribution to the debate will be discussed further in the next chapter, came to accept some role for group selection later in their careers.[36] In a review of Sober and Wilson's book, philosopher Elisabeth Lloyd wrote:

> In what is perhaps the most significant part of the book for practicing biologists, Sober and Wilson investigate the professional neglect of W. D. Hamilton's very early endorsement of the multilevel selection approach developed by G[eorge]

> Price. The multilevel selection models make it clear that genetic relatedness is only one of many ways to achieve the assortment into groups required by a process of selection at the group level. Especially helpful is the emphasis on the fact that game theory and kin selection models do not compete with multilevel Price-type models; rather, they are all different ways of looking at evolution in group structured populations.[37]

Lloyd went on to point out that much of the confusion resulted from the lack of distinction between the process of selection and the products of selection. I will present these distinctions in the next chapter.

In contrast to this moderate position that Williams described in the 1996 preface, a letter he had written to Lack in 1967 paints a more critical portrait. In a letter from Reykjavik, Iceland, in June he wrote: "I just recently finished your fine book on bird populations and would like to comment while the reading is still fresh. The reading was surely a pleasure. I had expected a nicely written and clearly reasoned account of a most interesting subject, and the book more than lived up to my expectations. Needless to say it has armed me with many valuable ideas." He continued to discuss some of the detailed examples presented in Lack's book that "raised difficulties for Wynne-Edwards' views, to put it mildly." Further down in the letter he lamented, "In a really reasonable world I should think that your discussion of Wynne-Edwards . . . would immediately settle the issues involved. But I guess that is too much to hope for." Finally, referring to his own experience he concluded:

> You probably had some trouble with the wording of your discussion of Wynne-Edwards. The subject requires great care to avoid the appearance of sarcasm or ridicule. I know that when I got to that part about the epideictic function of the vertical movement of plankton I suddenly wondered if I had fallen for a really elaborate joke. I believe that Braestrup's review indicated that people may react with laughter to a straightforward account of this theory. I had a similar but undoubtedly milder problem in discussing the ideas of Julian Huxley and others on evolutionary progress. One of Princeton's referees accused me of sarcasm and the use of straw men here, difficulties that I certainly wished to avoid.[38]

It is perhaps impossible to overestimate the effect of Williams's work on the field of evolutionary biology in general, and particularly on the theory of group selection and the fate of Wynne-Edwards's work.

Wynne-Edwards and the Ethologists

Beyond David Lack there was another Oxonian voice that is important to our story, that of the ethologist Niko Tinbergen. Tinbergen had come to Oxford after the war to establish a center for the study of animal behavior. In his 1981 paper "On the Emergence of Ethology as a Scientific Discipline," Richard Burkhardt wrote, "The history of scientific disciplines proves to be a history of social actions as well as scientific ideas."[39] The idea of discipline development as a social process of boundary establishment has been further expounded in the sociological literature, particularly in the work of Tom Gieryn.[40] In Burkhardt's 1981 article he pointed to the importance of Lorenz's assertion that "ethology, like the highly successful field of modern genetics, had come into being through the discovery and exploitation of a 'distinct and particulate physiological process.'"[41] I am especially interested in the idea of the ethologists, particularly Konrad Lorenz and Niko Tinbergen, that animal behavior and human behavior were equally appropriate subjects of biological analysis.[42] Lorenz's work emphasized the way special structures and behavior patterns had evolved in the service of intraspecific communication. Tinbergen's work centered on the survival value of various innate behaviors. Given Wynne-Edwards's focus on exactly these phenomena, one might expect their work to overlap more obviously. But Wynne-Edwards was interested in these behaviors purely in an evolutionary context, particularly the variation of responses to signals within a population. Lorenz interpreted behavior as a particulate trait useful for phylogenetic analysis in the same way that morphological traits had traditionally been used. Tinbergen focused on the fitness effects of particular behaviors on the individual. Wynne-Edwards, on the other hand, was preoccupied, almost to the exclusion of any other consideration, with explaining social behavior in an evolutionary context—that is to say, as an indicator of group fitness or a strategy for group survival. Lorenz, unlike the Skinnerian behaviorists he was setting his approach to behavior against, stressed the invariability of instinctive behavior but failed to attend to the phenomenon of intraspecific variation. Wynne-Edwards was expressly interested not in the conditioning of behavior in the psychological sense, but in the evolutionary importance of the group's ability to adjust behavior to given environmental conditions—the variability of groups. His emphasis on the evolutionary significance of group selection, combined with his lack of disciplinary identity, kept him outside the mainstream of the developing field of ethology and limited his role in its development.

Ultimately, Wynne-Edwards's influence on the development of ethology was similar to the effect he had on the related fields of behavioral ecology and sociobiology. Although his inclusion in Donald Dewsbury's 1985 collection of autobiographical memoirs, *Leaders in the Study of Animal Behavior,* might lead one to believe he was clearly identified as a behavioral biologist, my analysis shows that this is incorrect. I argue that clarifying Wynne-Edwards's relation to the field of ethology may help to define the development of the field.

Behavioral ecologist John Krebs described this influence succinctly in 1992 when he discussed the books that were most influential in his early career:

> I would like to start by admitting that the first of the three books that were very influential in the early stages of my career as a research scientist was one I did not read, although it was much talked about and vilified by some of the most influential teachers in my undergraduate course, Niko Tinbergen, David Lack and George Varley. The book in question was *Animal Dispersion in Relation to Social Behavior* by V. C. Wynne-Edwards. . . . I felt at the time, and still feel, that my results, along with other similar studies, vindicated Wynne-Edwards' view that behavioral mechanisms can be important in limiting population density on a local scale. This is not to say, however, that these mechanisms evolved by group selection.[43]

In detailing the relation between Wynne-Edwards's theory of group selection and the development of the field of ethology, I want to demonstrate how the response to this theory focused ethological work at the level of the individual, consistent with the rest of post–modern synthesis biology.

In September 1977 the Royal Geographical Society of London held a symposium on population control by social behavior. The stimulus for the symposium was Wynne-Edwards's theory of group selection, the idea that groups maintain population levels below environmental capacity through social organization, which replaces the direct, undisguised contest for food with competition for conventional prizes such as social status and territory, which in turn determine each individual's right to survive and reproduce. In the prologue to the proceedings F. J. Ebling wrote, "*Animal Dispersion in Relation to Social Behavior* consists of a comprehensive and elaborate documentation of all the data relevant to the hypothesis. Having first reviewed at length the visual, sound, electric, scent and tactile signals which are used for social integration throughout the animal kingdom, Wynne-Edwards proceeds to discuss the role of threat in establishing social hierarchies, whose "function (not hitherto defined) is always to distinguish the 'haves' from the 'have-nots' whenever the population density becomes excessive." This conclusion, Ebling continued, "was reached before the publication of Konrad Lorenz's book *On Aggression*; [in fact] Wynne-Edwards makes no reference to Lorenz (apart from a single paper written in 1931)."[44] Why is this noteworthy? In part, because their ideas appear quite similar; but also due to the popular interest in issues of population control that they both addressed. Ebling further noted,

> The attraction of Wynne-Edwards' hypothesis is undeniable. That over-exploitation of resources in the short term must be disastrous to the ultimate survival of animal and human societies alike seems axiomatic, and no intellectual effort is required to extend the concept of homeostasis, already familiar as applied to the internal environment of the organism, into the environment of the society. At the same time the hypothesis supplies a functional explanation for a range of behavioral phenomena in animals, particularly that known as ritualized aggression and given wide publicity by Konrad Lorenz (1966). With growing apprehension about the explosive growth of human populations it seemed, moreover, to point to a human moral, notwithstanding that proximate restraints on reproduction were invented by man long before the concept was considered in relation to animals. Indeed, one of Darwin's arguments for a "struggle for existence" was that animal reproduction was not subject to any "prudential restraint by marriage," which Malthus has previously recognized as a human attribute.[45]

Lorenz's and Tinbergen's stature within the biological community, culminating in their joint receipt, with Karl von Frisch, of the Nobel Prize for Medicine and Physiology in 1973, translated into major public recognition and discussion of their ideas; despite his focus on the same issues, Wynne-Edwards's work never reached the broader public to the same extent. In fact, Wynne-Edwards is best known as the father of a failed theory of group selection.

As Richard Burkhardt pointed out, the influential biologist Julian Huxley had a particular focus "on the benefit of the whole rather than the individual members of the species . . . that remained a common theme in his early papers."[46] The combined influence of Huxley and animal ecologist Charles Elton created in their student, Wynne-Edwards, a field biologist with an integrated approach to the study of animal behavior ecology.

As Gregg Mitman noted, "In the early 1920's [however] the mainstream of ecological research centered on either autecology, with its focus on the physiological response of individual organisms, or on community analysis."[47] Wynne-Edwards was educated and trained at Oxford in the midst of this division. Furthermore, Burkhardt notes, "What we now call ethology—that biologically oriented, comparative, and naturalistic approach to the study of behavior that we associate with Konrad Lorenz and Niko Tinbergen—did not begin to become a coherent enterprise until the 1930s and 40s, and even then its status was highly problematic."[48] Wynne-Edwards's staking out of social behavior as a mechanism for population homeostasis evolved through group selection creates an interesting comparison with the ethological approach to similar phenomena pursued by Lorenz and Tinbergen.

Wynne-Edwards's interest in behavioral features, such as the nonbreeding of sexually mature adults, that were not easily identified and that ran counter to Darwinian interpretation, precluded his inclusion in the developing field of ethology.

Through the 1930s and 1940s Wynne-Edwards pursued his ecological studies of behavior, focusing on the social structure of breeding populations. He was encouraged in this work by the developments in the modern synthesis that identified populations as the unit of evolutionary significance. The emphasis on higher-level selection was of obvious importance to Wynne-Edwards. As Gregg Mitman argued, "The past historical focus on such luminaries as Mayr and Simpson has given a false sense of closure to a number of issues in evolutionary biology that continued to circulate in the literature, especially where evolutionary thought intersected with ecology and behavior. . . . While most biologists by the 1940's believed natural selection to be the causative agent behind evolutionary change, still the question of what level(s) selection operated on remained a highly contested and unresolved point."[49]

Wynne-Edwards's interest in these very issues would place him in a difficult position vis-à-vis the developing ethological school.

But behavior per se was never made an integral part of the synthesis. As Burkhardt noted, "Huxley, in 1925, believed that the time had come to gather data from 'field observation, animal psychology & behavior, genetics, and comparative psychology . . . [and consider] the problem [of behavior] from a truly broad & unitary biological standpoint.' He failed, however to carry through on the project. What is more, in the broad, synthetic book that he eventually did write, *Evolution, the Modern Synthesis*, he neither made behavior a part of the synthesis nor offered guidelines to suggest how that might be accomplished."[50] Clearly, Wynne-Edwards thought his theory would accomplish the goal Huxley had set.

In 1959 Wynne-Edwards published his explanation for the control of population density through social behavior. He asserted in the opening paragraphs that most animals, because they can move, play a predominant role in their own population densities. Although he acknowledged that food was almost always the critical limiting factor, he maintained that birds are largely successful in regulating their own population densities below starvation levels through social conventions, or *epideictic* behavior. The term epideictic behavior applies to a very large class of social phenomena, which appear to have evolved for the primary purpose of demonstrating population density. Their purpose, according to Wynne-Edwards, is to supply the feedback into the homeostatic regulation of population density.[51]

As I discussed in chapter 4, in a 1959 article, the thrust of Wynne-Edwards's new hypothesis was illustrated by an analogy to human behavior. Citing man's own role as predator, he examined the problem of overfishing, arguing that "under a system of free enterprise, he is in grave danger of impairing the resource by taking too big a harvest, depleting the stock, and entering upon a spiral of diminishing returns."[52] He went on to point out that the way to avoid this situation is by an agreement binding participants to limit the catch to the long-term optimum figure. This analogy was generalized as a natural and inherent attribute of the relation between every kind of hunter and its staple prey. Wynne-Edwards continued to use the overfishing analogy throughout his published work (1962, 1968, 1971, 1978, 1986). This analogy, which imputes very complex intentionality to members of bird populations, aroused immediate suspicion on the part of Lorenz and especially Tinbergen, who were working to further "scientize" natural history and ethological fieldwork.

Konrad Lorenz

Lorenz became interested in evolution at age ten on seeing a picture of *Archeopteryx*, and he decided to become a paleontologist. As an undergraduate, studying medicine, he was quick to realize that comparative anatomy and embryology offered better access to the problems of evolution than paleontology did, but also that the comparative method was as "applicable to behavior patterns as it was to anatomical structure." It was also at this point that he realized that none of the people studying behavior really "knew" animals. Lorenz, who had kept all sorts of animals while growing up at his parents' home in Altenberg, Austria, took it as his challenge to develop this knowledge. Lorenz's most important early influence was Oskar Heinroth, whose studies of behavior served as his model. Lorenz was particularly impressed by Heinroth's comparative paper on Anatidae (waterfowl). It was during this period (mid-1920s to mid-1930s) that Lorenz developed his theory of instincts and worked on the concepts of releasing stimuli and vacuum activities. It was also at this point that Lorenz met Niko Tinbergen. He wrote in his autobiography, "At this Symposium (in Leiden in 1936) I met Niko Tinbergen and this was certainly the event which . . . brought the most important consequences to myself." The next year Tinbergen spent several months with Lorenz at Altenberg, where they conducted a series of experiments on egg rolling in the greylag goose. In 1939 Lorenz was appointed chair of psychology at Königsberg.

Lorenz was recruited into the German army as a medical corpsman in the fall of 1941. Six months later he was captured by the Russians and spent the next six years as a prisoner. In 1948 he returned to Austria and began looking for an academic position. Ultimately he established the Max-Planck-Institut für Verhaltenphysiologie (behavioral physiology) at Buldern. It was here that Lorenz would begin to train the first generation of ethologists. Lorenz's reunion with Tinbergen took place in 1949 in England, where Tinbergen had moved after the war.

In the later stage of his career, Lorenz was quite sympathetic to Wynne-Edwards's analogical approach, as evidenced by the title and topic of his Nobel Lecture, "Analogy as a Source of Knowledge." Here Lorenz argued, "The concepts of analogy and homology, are as applicable to characters of behavior as they are in those of morphology."[53] He went on to defend deducing function from behavioral analogies in this way: "Since we know that the behavior patterns of geese and men cannot possibly be homologous—the last common ancestors of birds and mammals were lowest reptiles with minute brains and certainly incapable of any complicated social behavior—and since we know that the improbability of coincidental similarity can only be expressed in astronomical numbers, we know for certain that it was a more or less identical survival value which caused jealousy behavior to evolve in birds as well as in man." This passage makes it quite clear that Lorenz shared Wynne-Edwards's conviction that analogy was an appropriate analytical tool for the ethologist. In the next section I will briefly discuss the relation between ethology and evolutionary theory from the perspective of Wynne-Edwards and Lorenz.

Ethograms

Although Lorenz often referred to behavioral traits that had evolved "for the good of the species," the hypothesis of group selection went beyond the purely descriptive behavioral categories, or ethograms, that he described as the basis for any further scientific study of behavior.[54] He wrote, "Above all, the first and indispensable prerequisite for permitting behavioral research to become a genuine inductive branch of natural science was missing: no more and no less than the idiographically and systematically assembled, *hypothesis-free* basis for induction."[55] Wynne-Edwards's analysis of behavior was clearly not "hypothesis free." He accepted Tinbergen's idea that fighting is conventional and provides hierarchy and territory, which benefit the individual, but he asserted that they are more importantly part of the homeostatic machinery of dispersion that maintains group fitness. Wynne-Edwards argued that sociality was the basis of conventional behavior and that together sociality and conventional behavior provided for the homeostatic control of dispersion. This, he claimed, was the first hypothesis that identified a common cause underlying all social behavior. Wynne-Edwards criticized Tinbergen for not seeing territory as "part of a larger and much more general phenomena—as just one among several categories into which we can conveniently group the systems of animal dispersion."[56]

Niko Tinbergen

Like both Wynne-Edwards and Lorenz, Niko Tinbergen grew up an avid amateur naturalist. He studied biology at Leiden University and developed his early interest in gulls. In 1925 he became an instructor in biology at Leiden, where he began his behavioral work on insects and birds. He wrote of his 1936 meeting with Lorenz: "Konrad's extraordinary vision and enthusiasm were supplemented and fertilized by my critical sense, my inclination to think his ideas through, and my irrepressible urge to check our 'hunches' by experimentation—a gift for which he has an almost childish admiration."

Tinbergen's collaboration with Lorenz was interrupted by the war. When Germany invaded Holland, Tinbergen was imprisoned in 1942 for resisting the nazification of Leiden University. After the war Tinbergen, in conjunction with Julian Huxley, David Lack, and William Thorpe, founded the journal *Behaviour,* which aimed at broader European collaboration in the study of behavior. In 1949 Tinbergen accepted a position at Oxford, where he was to become an influential promoter of ethology. (His students included Desmond Morris and Richard Dawkins, among many others.)

Although Tinbergen had been influenced by Huxley, in Burkhardt's words, "he did not follow Huxley in all of his enthusiasms. . . . In his own efforts to make the study of animal behavior more scientific, Tinbergen had stressed the importance of taking a more objectivistic and physiological approach."[57] Wynne-Edwards's suggestion that group-level selection was responsible for maintaining population density was exactly the kind of mushy thinking that Tinbergen was trying to relegate to the past. Wynne-Edwards's group explanation was also somewhat suspect from Lorenz's point of view. While Lorenz was clearly convinced of the importance of ritualization of competition, he perhaps was less sympathetic to the quasi-vitalistic nature of Wynne-Edwards's group selection theory, which seemed to impart some aspects of learning and animal mind that were not consistent with Lorenz's notion of innate behavior.

Wynne-Edwards also noted that his new theory was fraught with philosophical implications. This examination of the social behavior of animals, he claimed, provided the clearest indication yet of the "closeness of man's kinship with his fellow animals."[58] This invocation of the lessons we might learn about ourselves from this new theory not only echoes Darwin's claim (as a result of his theory) that "light [would] be thrown on the origin of man and his history,"[59] but also mirrors the idea of the ethologists, particularly Konrad Lorenz, that animal behavior and human behavior were equally appropriate subjects of biological analysis.

I introduce this connection to contrast Wynne-Edwards's explicit evolutionary approach (ultimate explanation) with the more physiological (proximate) approach of Lorenz. As Burkhardt argued, "The relations between ethology and evolutionary theory are not as straightforward as Lorenz might like. . . . Lorenz in particular emphasized the study of behavior and the way in which special structures and behavior patterns had evolved in the service of intraspecific communication."[60] Wynne-Edwards's published work since the 1950s, especially *Animal Dispersion,* addressed intraspecific communication extensively, and therefore one might expect Lorenz and Wynne-Edwards to overlap more obviously. But Wynne-Edwards was interested in these behaviors purely in an evolutionary [ecological] context, particularly the variation of responses to signals within a population. Individuals responded to signals based on their status as group members, for example, territory holder, dominant male. The attention Wynne-Edwards paid to nonbreeding behavior is a perfect example of this. The importance of this behavior, from his point of view, was precisely its variable expression or intraspecific variation. On the other hand, Burkhardt notes, "[Lorenz's] statements about species revealed virtually no engagement on his part with one of the key concepts of Darwin's thought, the concept of intraspecific variation. . . . Lorenz's understanding of what it meant for a science to be evolutionary was not formed by a direct reading of Darwin. Instead, it was mediated by that turn of the century emphasis on comparative anatomy."[61] Lorenz interpreted behavior as a particular trait useful for phylogenetic analysis in the same way morphological traits had been used. He emphasized the invariability of instinctive behavior and its mechanistic stimulation. Wynne-Edwards, on the other hand, was expressly interested in the evolutionary importance of the group's ability to adjust behavior to environmental conditions and in the behavioral variations among individuals of the same species, such as breeding and nonbreeding or differences in the onset of reproductive maturity.

This difference was also clearly influenced by the level at which each man's selective lens was focused. Consider the approaches of these two naturalists to aggressive and competitive behavior. Lorenz's preoccupation with ritualization, which accepts the virtues of aggression but looks for means to contain its vices, has approached the challenge largely from the viewpoint of the individual. Wynne-Edwards has taken the perspective of society. While both agree on the evolutionary value of competition, they disagree on the "innate" character of the organism. Aggression is a key element of each individual in Lorenz's community. Wynne-Edwards's organisms have evolved *not* innate aggression, but rather social behaviors that lead to population homeostasis and the avoidance of true aggression. Lorenz wrote in 1935 that "such co-operation of individuals in a colony is based entirely upon instinctive behavior patterns, just as in the case of the insects, and is nowhere based upon the traditionally acquired behavior patterns or upon the insight that co-operation in furthering the colony is advantageous to the individual."[62] This position was fundamental to Lorenz's and Tinbergen's explanation of behavior and allowed for the development of the studies of vacuum activities that Lorenz used to support his invocation of instinctual behavior. By contrast, Wynne-Edwards avoided the instinct concept altogether, because selection acted on a group property that could not be characterized as instinctual in any clear-cut way.

As Julian Huxley noted in his tribute to Lorenz in 1963, the advance in ethology that created the possibility for Wynne-Edwards to be considered influential was the development of the field by Niko Tinbergen. He praised "the light [Tinbergen's development] has shed on the origin of adaptive behavioural activities through the ritualization of conflict situations and their results, including displacement activities; and in particular, how selection has turned conflict and even aggression to useful account and utilized them in the construction of social bonds."[63] This characterization of Tinbergen's program might lead one to suspect that Tinbergen's and Wynne-Edwards's work might fruitfully cross-pollinate; but this did not happen.

Although Tinbergen wrote in the early 1960s that he was interested in studying "entire adaptive systems" and developed what he called a hierarchical system, the hierarchy in this case was a hierarchy of behaviors within each individual, as were the entire adaptive systems that he and his students at Oxford were examining. A passage from a paper Tinbergen presented in 1963 at the International Congress of Zoology makes this point explicit:

> In attacking our problem, let us start from observables—i.e., from behavior. But instead of studying its causes we shall study its effects; in other words, rather than look back in time, as we do when studying causation, we investigate what happens as a consequence of the observed behavior. . . . it differs from the study of behavior causation merely by the fact that the observed behavior is the cause, and that its effects are studied; we follow events with time instead of tracing preceding events, and we determine an animal's success. What I shall ask is simply, "Would the animal be less successful if it did not possess this behavior?" or, a little more subtly, "Would deviations from the perceived norm be penalized, and if so, how?"[64]

These are precisely the questions Wynne-Edwards attempted to address with his analysis of social behavior and his theory of group selection.

Indeed, in his introductory remarks Tinbergen went on to say that the proper approach to these questions was one that had become much maligned of late: "I shall approach this question as a naturalist—one who delights in observing animals in their natural surroundings."[65] Again, this seems to invite the kind of evidence gathering that made up the bulk of Wynne-Edwards's work on social behavior. Tinbergen went on to praise the importance of naturalist-style observation: "It cannot be stressed too much in this age of respect for—one might almost say adoration of—the experiment, that critical precise and systematic observation is a valuable and indispensable scientific procedure which we cannot afford to neglect."[66]

A deeper look at Tinbergen's 1963 paper shows his final verdict on Wynne-Edwards's interpretation of social behavior:

> [Wynne-Edwards's book] contains two main theses: first, many animals have developed means, usually behavioral, of preventing overcrowding; second, many of these means are "altruistic"—that is, beneficial to the population as a whole but not individuals—and as such can only be explained as consequences of group selection. While I believe Wynne-Edwards' first thesis to be sound—even though he seems to apply it to many phenomena that may well have other functions—his second thesis has the weakness of being based on negative evidence, on lack of analytical data.[67]

Tinbergen also argued that Wynne-Edwards's definition of altruism was too broad. He suggested that if it could be shown by concrete analysis that such forms of social interaction could arise as the result of conventional natural selection, such a theory, though of course not disproved in principle, would lose the only type of support Wynne-Edwards marshaled in its favor. And this is exactly what Tinbergen, Lack, and Williams proceeded to do. Indeed, in his memoir recalling his experience at Oxford, Robert Hinde wrote, "David Lack was no longer especially interested in behavior, but he taught me much about science, gave me a background interest in behavioral ecology, and made it difficult for me to ever think in other than individual selectionist terms—the sociobiological view that all ethologists were then group selectionists is nonsense."[68]

The commonalities of the perspectives of Lorenz and Wynne-Edwards persisted in their later work. Tinbergen's focus remained experimental and individual. He began behavioral research on early childhood autism and developed a center for child ethology at Oxford. Both Wynne-Edwards's and Konrad Lorenz's later work reflect a concern with the human population problem. Wynne-Edwards argued that the problem was with modern society. Humans no longer retained any of the natural adaptations that were effective in controlling population growth, although these adaptations remained functional in all the stone age peoples that survived into modern times.[69] Lorenz discussed the population problem at length in nearly all his popular works, from *On Aggression* through the posthumously published *On Life and Living,* where he stated, "All of the dangers that threaten mankind today result, in the final analysis, from overpopulation. None of them can be solved by any method other than education."[70] In the foreword to his 1987 work *The Waning of Humaneness,* Lorenz wrote: "Now, as never before, the prospects for a human future are exceptionally dismal. . . . Even if, just in time, humans should somehow impose a check on their blind and unbelievably stupid conduct, they still remain threatened by a progressive decline of all those attributes and attainments that constitute their humanity."[71]

Wynne-Edwards sounded equally dismal assessments. In his final published paper he wrote: "More than 10,000 years ago stone age culture began to give place to the agricultural revolution. . . . civilization was the revolution's early by-product. The moment people were released from dependence on a food territory and its natural produce, their basic group-selective mechanism, namely in-group selection, collapsed. . . . Nature's harsh measures have become unacceptable in our humane society, but fortunately for the biosphere, group selection is still at work on species living in the wild."[72]

The pronouncements of both men were consistent with contemporary concerns. Lorenz was never excommunicated from the biological establishment and therefore remained a strong influence in the development of animal behavior studies, and his books also continued to sell well. Wynne-Edwards, on the other hand, because of his dogged advocacy of a selective mechanism that was never accepted by the biological community, remained relatively obscure, outside the community of professional scholars interested in social behavior, population ecology, and evolution. Tinbergen worked on a number of popular films (for the BBC) and books (for Time Life) that presented ethology as the "biological study of behavior" to an increasingly broad audience.

In an analysis of the development of behavioral ecology, Peter Klopfer succinctly described the impact of Wynne-Edwards's work. After drawing connections from Allee's work on aggregation to Wynne-Edwards's presentation of social behavior and group selection as the proximate and ultimate causes, respectively, for population regulation, Klopfer presented the two concepts that became the dominant organizing principles in behavioral ecology: "First, optimality, which was introduced through Lack, MacArthur and by implication, Crook, Thorpe and Tinbergen; and secondly, game theory which grew from the work of W. D. Hamilton, G. Williams, Maynard-Smith and indirectly E. O. Wilson and R. Trivers. All of these workers were influenced by (if only by their reactions to) Wynne-Edwards."[73]

After these remarks regarding the influence of Wynne-Edwards, Klopfer added a somewhat startling observation, which raises questions regarding Tinbergen's and especially Lack's dismissal of Wynne-Edwards's theory.

> The notion of a stable structure-function relationship is solidly embedded in the biological corpus. It has a long history and is expressed in a variety of ways, particularly in post-Darwinian evolutionary biology, which, in turn, has had a pivotal influence on all of modern biology. But if we critically review the prevailing orthodoxies respecting the evolution and control of behavior, we often find that the experimental evidence that is claimed to support the traditional views is not compelling—that other conclusions are equally reasonable and compatible with the data.[74]

I suggest it is exactly this predisposition to individual-level thinking that led to the rejection of Wynne-Edwards's work by the ethological school and precluded a fuller integration of his behavioral work with that of Lorenz and Tinbergen.

Wynne-Edwards's advocacy for the theory of group selection had a deep but inconsistent effect on its development. On the one hand, it clearly stimulated a negative response from those committed to neo-Darwinian individual-level selection. On the other hand, Wynne-Edwards influenced some portion of the biological community to continue to examine the possibility that selection did indeed act on the group level and that individual-level explanations were not always sufficient. Both David Sloan Wilson and Michael Wade have acknowledged the influence of Wynne-Edwards on the early development of their thinking.[75] In both these cases, however, Wynne-Edwards's dogged commitment to his own formulation of group selection theory limited his influence and contribution to the debate as it developed through the 1970s and 1980s.

Arrested Development

What should we conclude regarding Wynne-Edwards's contribution to the development of ethology? I suggest that his influence on ethology in particular was much the same as his impact on biology in general. That is to say, while Wynne-Edwards's work is often cited, it is generally presented as a path that has not been fruitful for ethological researchers. As Mitman writes, "Once competition between individuals became the primary explanation of community structure, and individuality was defined by reference to genetic identity, the influence of group organization, of environmental context, on individual behavior became less significant."[76] Wynne-Edwards, despite his inclusion in Dewsbury's *Leaders in Animal Behavior,* was perhaps better described as an influential guidepost. His work forced ethologists to address the issue of behavior and population structure in an evolutionary context.

Although some might argue that Wynne-Edwards's unwavering commitment to his idea of group selection did more harm than good to the development of the theory, I think my analysis shows that this is incorrect. Wynne-Edwards may have diminished his own professional status as a result of his steadfast insistence on the importance of group selection, but he helped create the theoretical space where subsequent researchers could develop more careful analyses. His continued emphasis on the importance of understanding biological phenomena occurring above the level of the individual has been fundamental to the development of ecology and behavioral biology.

This commitment to individual-level adaptationist analyses of behavior has reared its head again in the newly developing field of evolutionary psychology. If we are aware of the history of the field of ethology, we can minimize the temptation to be satisfied by invoking oversimplified individual-level explanations and demand the perhaps more difficult but ultimately more evolutionarily accurate explanations provided by multilevel analyses.

Group Selection and the Human Population Bomb

In the late 1960s and early 1970s there was an increasing awareness of and interest in the environment. Rachel Carson's *Silent Spring* had raised public consciousness and concern regarding the environmental effects of industrial manufacturing and large-scale agriculture. Another component of this environmental awareness was increasing concern over human population growth. Indeed, Paul Ehrlich's 1968 book *The Population Bomb* precipitated massive public controversy and alarm, and once again Wynne-Edwards's idea of group selection would become enmeshed in the debate. In 1969 Garrett Hardin published his influential paper "The Tragedy of the Commons," which addressed local extinctions through the overuse of resources—exactly what Wynne-Edwards was proposing that group selection prevented. The function of group selection in population regulation was also a feature of Robert Ardrey's popular works *The Territorial Imperative* (1966) and *The Social Contract* (1970). In September 1971 he published in the *New York Times* two opinion columns, "Birth in the Wilds I & II," invoking the importance of Wynne-Edwards's theory of group selection. He began the first with the claim that "self-regulation of animal numbers is one of the more dramatic revelations made by students of animal behavior in recent decades." According to Ardrey, we have moved beyond Malthus to recognize that it is not an exhausted food supply that limits populations; rather, a "remarkable repertory of built-in mechanisms, behavioral or physiological, compel the normal species to keep the numbers of its young well within the carrying capacity of the environment." Unfortunately for humanity, we are among a small set of species (Ardrey's list includes lemmings, snowshoe hares, and humans) that have lost these mechanisms. He goes on to argue that this loss is the source of much human misery. In the second piece Ardrey discusses the positive effects of such modern ills as car accidents and drug addiction (as effective if not intentional means of population control), along with the importance of birth control and the value of homosexuals as nonreproductive members of human society. While the views expressed by Ardrey and Ehrlich are clearly meant to be provocative and extreme, many of the mainstream ecologists of the late 1960s and early 1970s were equally alarmed about the population problem.

A Blueprint for Survival

In January 1972 the new journal the *Ecologist* devoted the entire issue to an editorial titled "A Blueprint for Survival." The blueprint was written by the journal's editors, Edward Goldsmith, Robert Allen, Michael Allaby,

John Davoll, and Sam Lawrence, and outlined the environmental problems that the authors claimed governments were "either refusing to face . . . or briefing their scientists [about] in such a way that their seriousness is played down."[77] In addition to presenting the environmental problems they saw as most pressing (essentially pollution and overpopulation), they also presented the evidence they had used to construct their analysis and concluded with a statement of their goals. The details of their analysis and the precision (or lack thereof) of their predictions need not concern us here. The significance of this story is in the signing statement in support of the blueprint and the public controversy it precipitated.

Along with the tables and graphs that quantified the damage and made manifest the exponential increase in human population, the January issue of the *Ecologist* included a statement of support signed by many eminent members of the British ecological and environmental community. The list of signatories included the naturalists Sir Julian Huxley and Sir Frank Fraser Darling, the geneticists Douglas Falconer and C. H. Waddington, and not surprisingly, two of our main characters, David Lack and Vero Copner Wynne-Edwards. The list consisted of thirty-three supporters, many of them fellows of the Royal Societies of London and Edinburgh. Wynne-Edwards's signature here was particularly significant. Since the late 1950s he had served in a number of advisory and policy positions: he was a member of the Red Deer Commission from 1959 to 1968; he was president of the Scottish Marine Biological Association from 1967 to 1973; and most important, he was on the Natural Environment Research Council from 1965 to 1971 and served as its president from 1968 to 1971. In 1970 he was also appointed by the queen to the Royal Commission on Environmental Pollution. The commission was charged to "advise on matters, both national and international concerning the pollution of the environment; on the adequacy of research in this field; and the future possibilities of danger to the environment."[78] Given these affiliations, one might think that Wynne-Edwards would hesitate to sign what was sure to become a controversial document. But the temptation to apply his group selection idea to such a pressing social concern overrode whatever hesitation there might have been.

The response to the "Blueprint for Survival" in both the scientific and the popular press was quick. On January 14, 1972, the headline of the *New York Times* blared, "Britain Asked to Halve Her Population and Save the Environment to Survive." John Maddox, the editor of *Nature,* wrote a series of articles in April critically assessing the "doomsday scenario" described in the Blueprint. These editorials were subsequently satirized in a cartoon in the *Ecologist* depicting Maddox with an umbrella, standing waist-deep in water next to an ark and shouting "Alarmist!" up to the occupants of the ark.

Wynne-Edwards, of course, could not help but proffer group selection as the answer to the environmental and population threats humanity faced. In a paper he presented in Dallas, Texas, at the international symposium "Behavior and Environment: The Use of Space by Animals and Men," he reasserted some old themes: "I want to make it clear that a society is an organic entity in its own right; and especially that it has its own traditions, distinguishing it from other conspecific societies. Its biological function is to promote the welfare and survival of the stock that comprises it."[79]

According to Wynne-Edwards, this biological function of society is still operative in the animal societies, because for animals "toeing the line in the behavior code is always far more automatic." He continued, "Mankind abandoned the ancestral methods of achieving population homeostasis long ago, as part of the price of developing civilization. World population is tending more and more to become a single exploding group, plundering resources on a global scale."[80] He concluded the lecture in a vein consistent with the message presented in the Blueprint:

> It is because group selection depends so greatly on a pattern of space use that I have ventured to introduce it into this symposium. It has I believe played a leading part in evolution, not only in the development of social organization but of the physiological and genetic mechanisms as well. It has the important property of being able to select for adaptations that can only prove their worth in the long term, involving so many generations that the genes of any one individual and his family have become completely dispersed within the common pool. It can select for traditions which are the equal property of every member of a group, and have a large non-genetic component of inheritance which cannot be selected for through the fitness of the individual. Traditions likewise outlast the span of individual life, and influence the survival of the stock as a whole. But perhaps the most important function of group selection has been to find means of protecting the stock against the sabotage of short-term individual advantage.[81]

For Wynne-Edwards, then, the risk of signing the statement in support of the Blueprint was outweighed by the potential for revivifying his theory of group selection. He was also becoming more attuned to and explicit about the connection between group selection and population structure, which I will discuss in the next chapter. Indeed, a sentiment almost identical to the one voiced by Wynne-Edwards above appears in Sewall Wright's 1969 book *Evolution and the Genetics of Populations*: "Whether there is any way in which a species can be protected from parasitic genotypes that might arise within it or whether there is any way in which a species may evolve characters that are adaptive for it as a whole, although deleterious to the individuals that carry them, are important questions for evolutionary theory."[82]

CHAPTER SEVEN

The New Paradigm of the Gene

Genic Selection and Evolutionary Theory, 1964–1986

The influence of George C. Williams's book is difficult to overestimate. *Adaptation and Natural Selection* set a new standard for evolutionary studies and focused research at the level of the gene. The trend toward molecularizing biology had been under way for more than a decade, and Williams's sweeping critique of group selection theory marked the beginning of a new age of genic selection. Its basic idea was simply that natural selection is always, or for the most part, selection for and against single genes.[1] Of course this position was not universally accepted. Ernst Mayr had maintained since the early 1960s that although natural selection had an effect on gene frequencies, it was not necessarily selection for or against particular genes. In *Animal Species and Evolution* Mayr stated his position unequivocally: "Natural Selection favors (or discriminates against) phenotypes, not genes or genotypes. Where genotypic differences do not express themselves in the phenotype (for instance, in the case of concealed recessives), such differences are inaccessible to selection and consequently irrelevant."[2]

Despite this rather clear statement of the problem with the genic selectionist account, the paradigm continued to gain influence throughout the 1960s and 1970s.

William D. Hamilton and John Maynard Smith

The decade following Lack's and Williams's critiques was one of continual decline in the fortunes of Wynne-Edwards and of group selection. William Hamilton, a theoretical biologist from Oxford who had devised the theory of kin selection in 1964, continued to hone his ideas and drive the focus of evolutionary biology to the level of the gene.[3] John Maynard Smith, a student of J. B. S. Haldane, one of the founders of population genetics, also engaged in an ongoing criticism of Wynne-Edwards's work and supplied perhaps the most crucial challenge to his theory: the possibility of invasion of an altruistic population by an opportunistic cheater. In 1971 Robert Trivers published his paper on reciprocal altruism, which again thwarted the population regulation explanations that Wynne-Edwards had proffered for various seemingly self-sacrificial behaviors.[4] The culmination of the gene-centered view came with the 1976 publication of Richard Dawkins's *The Selfish Gene*. Wynne-Edwards occasionally responded to these developments in print, as in his exchanges with Maynard Smith in the pages of *Nature*, on the difference between group selection and kin selection, and with Christopher Perrins and David Lack on the differential survival of swifts in relation to brood size. Maynard Smith argued,

> Both genetical theory and the experimental evidence suggest that if natural selection has been pushing a character in a given direction for a long time, it will be difficult for selection to produce further change in the same direction, but comparatively easy to produce a change in the reverse direction. Thus it would only be plausible to suggest that there are genetic reasons why anti-social behavior should not increase if it were also suggested that selection had already produced an extreme degree of anti-social behavior, and this is precisely what Wynne-Edwards denies. In fact, "anti-social" mutations will occur, and any plausible model of group selection must explain why they do not spread.[5]

Having issued this challenge to Wynne-Edwards, Maynard Smith used the rest of his communiqué to develop his haystack model of group selection, which he used to demonstrate that kin selection was an excellent alternative to group selection. Like the proposals of J. B. S. Haldane and Sewall Wright, the haystack model required groups that were spatially isolated multigenerational units.

First, Maynard Smith supposed a species of mouse that lived entirely in haystacks. An individual haystack would be colonized by a single fertilized female, whose offspring form a colony that lives in the haystack until the next year, when new haystacks would become available for colonization. Then the mice migrate and possibly mate with members of other colonies before establishing a new colony. The population would consist of aggressive A and timid a individuals (timidity would be caused by a single Mendelian recessive gene, so only homozygous aa individuals would be timid, and all others would be aggressive). It followed then that only colonies started by homozygous recessive females fertilized by homozygous recessive males would produce a colony of timid individuals; all other colonies would lose the a gene by selection and become entirely populated by A individuals. It is assumed on this model that the a populations would contribute slightly more individuals to the mating pool and would have a proportionately greater chance of having a daughter colony. Thus, with little or no interbreeding between colonies even at migration, timid behavior would evolve, assuming it is advantageous to the group. However, according to Maynard Smith this conclusion is rather unimportant, because this kind of division into a large number of completely isolated small groups is highly unlikely.[6]

For group selection to work on this model, altruism first had to be established in some of the groups by genetic drift. This is the first highly unlikely event that is not required by the kin selection model. Second, as described above, the populations in which altruistic behavior could be established would have to conform to a set of constraints not often observed in nature.

Wynne-Edwards's response to Maynard Smith's model is particularly interesting. First, he pointed out that the differences in approach were partially a result of their respective backgrounds, a laboratory geneticist versus a field ecologist. "To me his picture of territorial systems and other aspects of conventional behaviour appears scarcely true or comprehensive enough to provide a basis for valid deduction; my own grasp of the genetical theory of natural selection, on the other hand, no doubt looks still more halting and inept to him."[7]

Nevertheless, Wynne-Edwards went on to provide some further criticism of Maynard Smith's model regarding population structure, which proved quite correct. He pointed out that most ecologists "would agree that the prerequisite of group selection that calls for a subdivided population structure is commonly and indeed normally found in animals."[8]

Finally, Wynne-Edwards argued that the model of the haystack was not a sufficiently close approximation to any natural situation to help move toward a solution. "A realistic counterpart might be, for example, the woodlice (*Porcellio scaber*) that fed on the green alga *Protococcus* living on tree trunks studied by Brereton; marked woodlice confined their feeding to their own particular tree, and the population appeared to be subdivided into breeding units. Had any of the latter increased too freely they could have exterminated their stock of this particular food plant which does not regenerate easily."[9]

Despite this counterexample, the haystack model became the standard example of group selection, especially for those opposed to it. In a 1989 paper, Gregory Pollock pointed out that the success of Maynard Smith's model was built on a conception of the group that was not consistent with what Wynne-Edwards had argued. According to Pollock, the haystack model provided the basis for rejecting Wynne-Edwards:

> [In the haystack model] groups were assigned the same indivisible integrity as individuals. Conjectured group advantageous traits such as altruism (i.e. foregoing some degree of personal reproduction to aid others) were, by definition, expressed homogeneously within groups; intra-group heterogeneity was always eliminated through intra-group selection. The appearance of a single cheater, foregoing altruism for an immediate reproductive advantage in its group, then necessarily converts a group into a collection of selfish individuals. Group selection would require that groups with cheaters be eliminated as fast as they appear; that is, that the rate of group extinction is identical to the individual mutation rate for cheating, a not quite absurd possibility in nature.
>
> Wynne-Edwards viewed populations as often subdivided into partially isolated groups in that individuals frequently remain near their birthplace to reproduce in turn ("philopatry"). Maynard-Smith made a crucial and, for Wynne-Edwards, fatal distinction by arguing that such population viscosity tends to aggregate relatives; selection for altruism in such cases is then a product of kin selection via Hamilton's inclusive fitness theory and not group selection, as the former does not require "partially isolated breeding groups" with some form of the group phenotypic integrity, as in a shared group trait for altruism.[10]

Pollock's point is well taken. Indeed, if Wynne-Edwards's theory of group selection is understood without the assumptions presented as necessary by Maynard Smith, it is compatible with more recent models of group selection. In these models groups are not defined in terms of spatial isolation, as in the haystack model, but represent behavioral isolation for an incomplete portion of the life cycle. These comments should not be interpreted to mean that Wynne-Edwards's theory of group selection had gotten it right. Rather, Pollock's analysis (consistent with his experience in Michael Wade's laboratory) shows that the haystack model misrepresented Wynne-Edwards's theory of group selection and provided an easy target for critics.[11] Wade's 1980 paper in *Evolution* gave an experimental test of Maynard Smith's ideas, and a second paper in *Science* formally partitioned kin selection into within-group and between-group components and showed that Hamilton's rule was equivalent to saying that group selection was stronger than opposing individual selection.[12] Indeed, Wade's experiments found that selection with "partially isolated groups" (group selection) was much more efficient than selection in randomly mating groups (kin selection). That is, these results did not support Maynard Smith's arguments.

Another criticism of the haystack model, this one provided by David Sloan Wilson and Elliott Sober, identifies two counts on which Maynard Smith's distinction between group selection and kin selection can be faulted. First, he failed to recognize that multigroup population structure exists during the first generation, when the original siblings interact with each other. The processes of within-group and between-group selection that occur after the first generation in the haystack model are merely a continuation of those that occur during the first generation. Second, even if we accept Maynard Smith's definition of group selection, his pessimistic conclusion is based on the assumption that within-group selection is as strong as it can possibly be. If we use Hamilton's or Wright's equations for altruism in the haystack model, then altruism can evolve by group selection even as defined by Maynard Smith.[13] Maynard Smith's model implicitly allowed very strong selection within groups to oppose altruistic behavior, because he assumed that all groups founded with even a single selfish allele would become completely selfish earlier than he allowed for it to happen under group selection between completely altruistic and completely selfish groups. As a result, any migration of selfish genes into an altruistic group eliminated the altruistic character of the group and converted it into a wholly selfish group. Without strong isolation, altruistic groups could not persist.

Later, genetic criticisms of Maynard Smith's haystack model were to emerge (Pollock's in 1989 and Sober and Wilson's in 1998), but during the 1960s and 1970s the haystack model was incredibly influential in discounting the possibility of group selection. The combination of kin selection and Maynard Smith's model became the paradigm for evolutionary biologists interested in social behavior. Given the portrait of Wynne-Edwards's tenacity and single-mindedness to this point, it is not surprising that he remained resolute in his beliefs. In his personal notebook from the late 1960s, we can see an attempt to mold Hamilton's thoughts more toward his own. After reading Hamilton's two 1964 papers, he wrote: "Does Hamilton's idea of 'inclusive fitness' wh. takes account of replica genes in a person's relatives, not perhaps apply to whole gene pools. Is individual fitness all a fallacy, based on erroneous Darwinian assumptions of 'striving to increase in numbers'? The surviving organism is the population. Individs come under selection & this removes bad genotypes (bad combinations). Selection as a cleansing process."[14]

A series of comments follow this initial analysis and focus on the problems of Hamilton's theory. Wynne-Edwards noted, consistent with his line of thought from the mid-1950s, that it was inadequate to explain the evolution of social adaptations based on individual advantage as Hamilton had done. Given that the benefit did not accrue to the individual, it was not possible for selection acting on the individual to produce this group result. This line of thought was followed in the notebook by some passages from Dobzhansky's 1951 article, "Mendelian Populations and Their Evolution."[15] The important passage from Dobzhansky that Wynne-Edwards copied out asserted, "In other words, natural selection enhances the adaptedness of the Mendelian popn as a whole, at the price of continuous produc.n of some less well adapted individs."[16]

This passage connected some of the most recent theorizing against Wynne-Edwards's theory (inclusive fitness theory) with the earlier thinking of one of the architects of the modern synthesis. Wynne-Edwards interpreted Dobzhansky's work as contradictory to the idea of kin selection, because the populations he described were not necessarily composed of closely related individuals. The "benefit" of genetic variation could not be ascribed to the individual because it was a characteristic of the population.

The connection to Dobzhansky's theorizing was enhanced by a personal connection the two made in 1966 at a conference at Rockefeller University on the relation of biology and behavior. Dobzhansky gave the keynote address, "Genetics and the Social Sciences," and Wynne-Edwards gave a paper on population size and social selection.[17] According to Wynne-Edwards's recollection in his 1985 autobiography, it was at this meeting that Dobzhansky offered a challenge:

> In 1966 I took part in a conference at the Rockefeller University in New York on genetics and the social sciences. The distinguished geneticist Theodosius Dobzhansky was the leading speaker, and in private afterward, while we were discussing the implications of my own paper on population control and social behavior in animals, he put a specific challenge to me. "Don't you think you could sort this thing out once and for all?" he said. "This thing" was how group selection (which has vexed geneticists for half a century) could take place, and lead to the evolution of collaboration and altruism, and of statistical kinds of genetic adaptation as well.[18]

The subsequent paragraph of Wynne-Edwards's autobiography suggested that he might have finally met Dobzhansky's challenge in his forthcoming book, albeit too late for Dobzhansky to witness. The completion and reception of Wynne-Edwards's second book-length treatment of group selection will be presented in the second half of this chapter. Before discussing the 1986 book I will examine two more of the major theorists fundamental to what Elliott Sober and David Sloan Wilson, in *Unto Others*, have dubbed the "dark ages of group selection."

Edward O. Wilson and Richard Dawkins

In 1975 the Harvard entomologist Edward O. Wilson published his new theory of animal behavior in a sizable volume titled *Sociobiology: The New Synthesis*. This work represented Wilson's attempt to understand all of animal behavior (ultimately including human behavior) in terms of evolutionary adaptiveness. *Sociobiology* created a storm of contention almost immediately. The sociobiology controversy was perhaps the most publicly debated episode in the field of biology since the Scopes trial in Dayton, Tennessee, in 1925. Although Wilson had many supporters, he also suffered a great deal of criticism from within the scientific community and without. Biologists criticized Wilson's methodology and adaptationist reasoning, philosophers of science challenged his attempt to "biologicize ethics," and sociologists, educators, feminists, and others deplored his apparent ignorance of the social and political implications of his work.[19] In his 1994 autobiography Wilson wrote that the reviews of *Sociobiology* "whipsawed it with alternating praise and condemnation."[20] Along with the wide-ranging response to his theory, Wilson also recalled something about his methodology: "In order to use models of population genetics as a more effective mode of elementary analysis, I conjectured that there might be single, still unidentified genes affecting aggression, altruism, and other behaviors."[21]

Wilson's sympathy for the gene's-eye view is apparent in the passage above. His commitment to genic selection was such that in the fifth chapter of *Sociobiology* he rejected most of Wynne-Edwards's argument. He argued, essentially, "that one after another of Wynne-Edwards' propositions about specific 'conventions' and epideictic displays were knocked down on evidential grounds or *at least matched with competing hypotheses of equal plausibility drawn from models of individual selection*."[22] Note that Wilson's rejection of Wynne-Edwards's theory does not differ much from David Lack's of a decade earlier, which was carefully reviewed in chapter 6.

Despite this focus on the gene, Wilson's work was important to Wynne-Edwards in several ways. First, Wilson was not only an entomologist but more specifically a myrmecologist. As I described in earlier chapters, the social insects had been, and continue to be, of major importance to the development of thought regarding higher-level selection and social organization. Second, Wynne-Edwards felt a commonality with Wilson. They were both witnesses to the transition from the traditional naturalist to the modern ecologist and committed to the global theorizing that the younger generation avoided.[23] Finally, Wilson's work on sociobiology provided a foil to Wynne-Edwards's own generalist approach. Whereas the criticisms and examples given by Lack were generally restricted to avian biology, in Wilson Wynne-Edwards had found a neo-Darwinian willing to address selection theory in broader applications. Wynne-Edwards initially hoped that a work like Wilson's *Sociobiology* might renew interest in his theory of group selection. Given Wilson's commitment to explaining social behavior in terms of kin selection theory, however, group selection continued to be ignored.

In a review of Scott A. Boorman and Paul R. Levitt's *Genetics of Altruism*, Wynne-Edwards pointed out that Wilson's discipline of sociobiology had a fundamental defect.

> Because [Wilson] could not accept its reliance on group selection he dismissed my hypothesis, with its definition of society as "an organization capable of providing conventional competition among its members," and its concept of social life and population homeostasis evolving together as parts of a functional whole. . . . For Wilson, a society is "a group of individuals belonging to the same species and organized in a cooperative manner." He attaches no significance to conventional competition or conserving resources, and not much to population regulation. His key properties of social existence, including cohesiveness, altruism, and cooperativeness, are sufficiently imprecise for him to suggest that some of the primitive colonial invertebrates, such as corals and siphonophores, come the closest of all animals to producing perfect societies. . . . In the rest of the animal kingdom the groups he picks out as social are generally the ones conspicuous for what one might call sociability, that is gregariousness; and in tables showing the incidence of sociobiological traits in the different phyla, "social" animals are explicitly contrasted with "solitary" ones. . . . In short, lacking a valid definition of society, Wilson's synthesis sometimes sheds more confusion than light.[24]

The comments in this review illuminate why Wynne-Edwards ultimately rejected Wilson's theory and vice versa. The idea that social conventions could be explained in terms of individual selection was completely contrary to Wynne-Edwards's view of nature. Indeed, he remained convinced that Wilson (among others) lacked a basic understanding of sociality. On the other hand, Wilson's commitment to genic-level explanation precluded the consideration that groups might act as units of selection. In her account of the sociobiology debate, sociologist Ullica Segerstråle points out that Wilson was a bit unclear on this point and that it was only through Robert Trivers's revisions in one of the last drafts before publication that Wilson sorted out his position.[25]

The renewed interest that Wilson's work had brought to the study of the social behavior of animals could have been a great benefit to Wynne-Edwards's project; but with the emphasis of the sociobiologist on the natural selection of particular genes, Wynne-Edwards saw that group selection was considered less and less important in evolutionary theory.

Perhaps the most fascinating development here is that Wilson has recently revised his position with respect to group selection. In a 2005 paper written with his longtime collaborator Bert Hölldobler, he now argued that "group selection is the strong binding force in eusocial evolution."[26] Indeed, more even more recently Wilson wrote a paper with David Sloan

Wilson, the foremost contemporary group selectionist, titled "Rethinking the Theoretical Foundation of Sociobiology." In the abstract to this paper the authors pointed out that "current sociobiology is in disarray. . . . Part of the problem," they continued, "is a reluctance to revisit the pivotal events that took place during the 1960s, including the rejection of group selection and the development of alternative theoretical frameworks to explain the evolution of cooperative and altruistic behaviors."[27] I will discuss this change of heart a bit more in the concluding chapter.

Throughout the 1960s and 1970s Wynne-Edwards maintained a series of notebooks in which he prepared himself for a second pass at group selection. Interspersed with his notes on the literature and his synopses of contemporary publications in evolutionary theory, Wynne-Edwards would write short passages addressing the themes that gave him the most concern. Many of these passages became the subject matter of his various lectures and papers throughout the 1970s. For example on May 16, 1975, he focused on altruism. "In man," he wrote, "the spiritual rewards of public service include the satisfaction of (a) teaching the young (b) ministering to the sick (c) comforting the distressed & c. People devote their careers to such work & not all are well rewarded financially. Typically they are altruistically motivated & dedicated (? Does devotion to 'prayer & fasting' bring spiritual reward? It has not much adaptive value)."[28]

The year after Wilson's publication, the British ethologist Richard Dawkins's *The Selfish Gene* arrived on the scene. Dawkins took the gene-centered point of view and presented it very persuasively to both popular and professional audiences. In the preface to the 1989 revised edition, he wrote of his personal goal, "I would write a book extolling the gene's-eye view of evolution. It should concentrate its examples on social behaviour, to help correct the unconscious group-selectionism that then pervaded popular Darwinism."[29] In Dawkins's recollection, the intellectual bogeyman of group selection was still prevalent, at least when he began writing his book in 1972. He went on to describe *The Selfish Gene* as merely the logical outgrowth of the orthodox neo-Darwinism of R. A. Fisher and perhaps a more "full-throated account" to accentuate the "laconic expressions" of George C. Williams and William D. Hamilton that had so inspired him.[30] Dawkins further noted that the increasing focus on individual and genic-level selection had led Wynne-Edwards in 1977 to what Dawkins described as a "magnanimous recantation" regarding group selection. Wynne-Edwards wrote, "In the last 15 years many theoreticians have wrestled with [group selection] and in particular with the specific problem of the evolution of altruism. The general consensus of theoretical biologists at present is that credible models cannot be devised, by which the slow march of group selection could overtake the much faster spread of selfish genes that bring gains in individual fitness. I therefore accept their opinion."[31] The recantation was short-lived, as Dawkins described in the revised edition of *The Selfish Gene:* "Magnanimous these second thoughts may have been, but unfortunately, he has had third ones: his latest book re-recants."[32]

Although there is no published account of Wynne-Edwards's reaction to Dawkins, or any correspondence between them, his notebooks again fill out the narrative. In notes written in October 1978 Wynne-Edwards laid out several objections to Dawkins's account.

> Single genes (v. stable chemically) cannot acquire adaptns thru nat. seln. & thus evolve. DNA exists in millions of forms, all equally [illegible] in that they have survived and are surviving. When genes mutate. "rival" alleles are created.
>
> Essential test of evolving in acquisition of favourite hereditary adaptations thru nat seln, & it belongs to the species consisting of mortal individuals (species and subdivisions down to demes) (i.e. gene-pools evolve rather than genes).
>
> If genes pursued self-promotion, success would be attained by the fixation of loci, so that genes passed to 100% of offspring. Instead we find (1) mutation at rates that are heritable and genetic, (2) heterosis which prevents alleles from competing.
>
> Genes for asexual reproduction or self-fertilising would always win against those for heterosexuality.[33]

These notes are included here not so much for their probing into Dawkins's faults as for the illumination they shed on the development of Wynne-Edwards's work. The three points made above lay the foundation for the basic question of the entire group selection debate. Which is the more robust evolutionary approach? Is it the static gene or the hierarchical and developmental aspect? They also serve as an interesting example. During this period Wynne-Edwards was already working on the outline for his second book. Despite considering competing hypotheses in his notes, he almost never acknowledged or attempted to refute them in print—even in the second book. In the broader context, *The Selfish Gene* served as the nail in group selection's coffin, built ten years earlier by George C. Williams's *Adaptation and Natural Selection.*

Philosophy of Biology

The issues raised by the controversy over group selection, which developed over time into the "units of selection debate" or the "levels of selection question," were not straightforwardly empirical. This is evinced by the immense quantity of philosophical ink spilled over the topic. Since Richard Lewontin's 1970 article on the units of selection, in which he argued that all entities that exhibit heritable variance in fitness are in fact units of selection, philosophers have focused on this issue. Varied efforts toward terminological clarification and the analysis of more fundamental questions, such as the proper identification of biological individuals or the privileged position afforded to lower levels of explanation by those sympathetic to the gene-centered point of view advocated by George C. Williams and Richard Dawkins, have been addressed and debated.[34] Other philosophers and philosophically inclined biologists involved who contributed to the debate include David Hull, who wrote several articles examining the role of the individual and its ontological status in evolutionary theory,[35] and William Wimsatt, who wrote multiple articles on the effectiveness of reductionistic research strategies and their role in the developing units of selection debate.[36] Philosophers Robert Brandon, Kim Sterelny, Philip Kitcher, Michael Ruse, and Richard Burian also contributed.[37] Philosophers and biologists have also collaborated. Elliott Sober, who has written some the most important philosophical work on the levels of selection, has worked with Richard Lewontin and more often with the population biologist David Sloan Wilson.[38] Elisabeth Lloyd, whose 1988 book *The Structure and Confirmation of Evolutionary Theory* contributed a great deal to untangling many of the fundamental issues in the units of selection debate, has worked with Stephen Jay Gould on important articles on species selection.[39] The biologist Michael Wade, whose laboratory investigations of group selection have been fundamental to the contemporary understanding of the theory, has written papers with philosopher James Griesemer.[40]

In a recent paper, Lloyd has provided a very useful anatomy of the units of selection debate that identifies four questions being asked as part of the this debate and assigns the major players to their appropriate questions.[41] First, What units are being actively selected? Are they genes, genomes, organisms, groups, or species? This is known as the interactor question. The second question is What is the replicator? Here Lloyd follows David

Hull's definition of a replicator as "an entity that passes on its structure directly in replication." The third question asks, Who (or what) benefits from a process of evolution by selection? Lloyd points out that the beneficiary question permits two interpretations. The first version is Who ultimately benefits from the evolutionary process? This connotes a long-term evolutionary beneficiary, and the second version asks, Who benefits from the ownership of the adaptation? This indicates a short-term-adapted beneficiary. This distinction leads to Lloyd's fourth question, which embraces the levels of selection debate: At what level do adaptations occur? Or as she cites Elliott Sober, "When a population evolves by natural selection, what, if anything is the entity that does the adapting?"[42] In the rest of the article Lloyd uses this set of questions to analyze the positions of the major figures in the units of selection debate. Although I will not recapitulate her analysis here, I introduce the questions to show the complexity of the debate and to provide some idea of the volume of philosophical literature that has been generated in attempts to settle these questions. Furthermore, the first half of this chapter provides the intellectual environment that Wynne-Edwards would have to adapt to in writing his second book-length treatment of group selection, *Evolution through Group Selection.*

This was to be his definitive account of the power of group selection in shaping evolutionary history. Much of the book was devoted to an interpretation of the long-term research on the red grouse that had been conducted at the Culterty Field Station, which Wynne-Edwards had worked to establish in the late 1950s. The grouse research would stand as the detailed example or case study of the role of group selection in maintaining population levels that many of Wynne-Edwards's critics had demanded since the publication of *Animal Dispersion.* Equally important to Wynne-Edwards's strategy to establish the fundamental importance of group selection was incorporating Sewall Wright's shifting balance model into his own work.

Making Group Selection Wright

Wynne-Edwards's notebooks demonstrate an increasing interest in Wright throughout the 1970s that culminated in the 1986 book, which included a chapter on evolution in structured populations that drew heavily on Wright's shifting balance model. Wynne-Edwards stated at the beginning of the chapter, "On the genetical side, however, it appears that the classical assumptions about panmixis in mating systems are seldom if ever borne out by the facts, at least in the vertebrates; and in birds and mammals including primitive races of man, mating is often regulated and contrived—often as in the red grouse being exogamous and patri- or matrilocal as the case may be."[43] Wynne-Edwards went on to discuss how structuring of populations could readily restrict gene flow between groups or demes. Population structure performed two important functions, one ecological, the other evolutionary. The ecological function, to ensure that the rewards of resource husbandry were transmitted to subsequent generations, had long been the focus of Wynne-Edwards's work on group selection. In *Evolution through Group Selection* he had finally turned explicitly to the evolutionary function of population structure, and that turn led directly to Wright's shifting balance theory. He wrote: "My own ideas about both functions [ecological and evolutionary] have of course been reached by following ecological and sociobiological paths; but there is an older and more powerfully and persistently argued theory that has led to very much the same conclusions, with regard to the evolutionary conclusions alone, in the work of Sewall Wright (e.g. 1931, 1940, 1945, 1949, 1968–78, 1980)."[44]

He emphasized that Wright's argument was based on the usual premise that evolutionary change is mediated by changes in gene frequency and that these changes are due to mutation, selection, migration, and chance. He was particularly interested in the characters of organisms that exhibited continuous variation. Since these characters are largely polygenic, they provided the variation of fine adjustment, and for Wynne-Edwards they represented the main systems of demic differentiation. In structured populations, then, selection and chance were the most powerful evolutionary forces. "In such small units the individuals that actually survive to breed will be rather few in number, and subject therefore to appreciable sampling errors, giving rise to corresponding variances between units even though the units may have been originally derived from the same base population."[45] Wynne-Edwards went on to describe Wright's adaptive landscape and included the now famous illustration from his 1932 paper.

Wynne-Edwards then cited Wright's 1940 paper, where he quoted,

> "Under this condition neither does random differentiation proceed to fixation, nor adaptive differentiation to equilibrium, but each local population is kept in a state of continual change. A local population that happens to arrive at a genotype that is peculiarly favorable in relation to the general conditions of life of the species, i.e. a higher peak combination than that about which the species

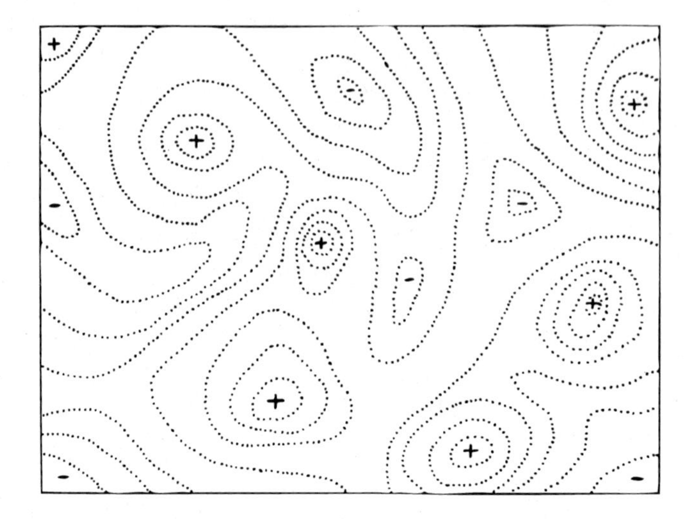

FIGURE 14. Wynne-Edwards expanded his discussion of Wright's ideas in his second book on group selection and included Wright's adaptive landscape diagram above to demonstrate the consistency of their ideas. From Sewall Wright, "The Roles of Mutation, Inbreeding, Crossbreeding and Selection in Evolution," *Proceedings of the Sixth International Congress of Genetics* 1 (1932): 356–366.

had hitherto been centered, will tend to increase in numbers and supply more than its share of migrants to other regions, thus grading them up to the same type by a process that may be described as intergroup selection" (Wright 1940 p 175).[46]

This introduction of Wright's focus on population structure, culminating in his description of intergroup selection, led to Wynne-Edwards's reading Wright's 1980 paper "Genic and Organismic Selection." This paper came to Wynne-Edwards like manna from heaven. He began his analysis by pointing out that, much like his own, Wright's theory had frequently been misrepresented, and that the paper was meant to set things straight. This indeed was the same motivation behind Wynne-Edwards's second book. What he found particularly fascinating was that their ideas were so similar and that they had been "misunderstood" in similar ways. Wynne-Edwards argued,

Populations structured on Wright's model could store a much greater genetic variability than panmictic populations of the same size; and not only might better adapted stocks emerge in the manner just described, but when the environmental changes of wider than local importance occurred the probability would be correspondingly greater that preadapted units were already in existence, from which mass selection and expansion could take place. In his expressive phrase (Wright 1980:841) "creativity is raised to the second power" in populations structured in this way.[47]

Here again we see Wynne-Edwards homing in on Wright's insights regarding population structure and trying to connect them with his own. In Wynne-Edwards's eyes this was perhaps the most important function for the 1980 paper. In Wright's attempts to clarify his ideas to the community of evolutionary biologists, Wynne-Edwards saw an opportunity to show the consistency of his own thought with Wright's and thereby salvage his theory from oblivion. He was particularly keen to show that his theory, like Wright's, need not always be considered as working in opposition to individual-level selection. Given the earlier critiques of Williams and Maynard Smith, this was an especially important point of clarification for him. He cited Wright gleefully:

[Wright] concludes the paper by saying that several recent authors who have discussed group selection for group advantage at length, have rejected it as of little or no evolutionary significance, and seem to have concluded that natural selection is practically wholly genic. "None of them discussed group selection for organismic advantage to individuals, the dynamic factor in the shifting balance process. . . . although it is not fragile at all, in contrast with the fairly obvious fragility of group selection for group advantage, which they considered worthy of extensive discussion before rejection." My task is obviously to show that the group selection I advocate is not essentially different from Wright's in raising average individual fitness, and is not fragile either.[48]

Wynne-Edwards's extensive reliance on Wright in *Evolution through Group Selection* was preceded by a decade of reading and annotating Wright's work and connecting it to the work of Dobzhansky and the Chicago ecologist Warder Clyde Allee, whom Wright had worked with through the middle part of his career. Indeed, while working on the manuscript for the book, Wynne-Edwards had corresponded with Wright regarding Allee. Wynne-Edwards was interested to hear Wright's thoughts on Allee's use of Wright's population structure models as a functional explanation of undercrowding in laboratory populations of invertebrates. He wrote to Wright that this was a "remarkable stroke of intuition" and that he had recently come to the same conclusion himself. Wright responded that he was certain he and Allee had discussed these matters at the time, but he couldn't recall the nature of the conversations beyond what had been published in *Principles of Ecology*. He went on,

> I was interested in learning that you are writing a sequel to your 1962 book. I touched on the subject of the conditions under which selection for group advantage would prevail over an associated individual disadvantage in a review of [George] Simpson's *Tempo and Mode of [sic] Evolution*. I overlooked then a discussion by Haldane in his 1932 book on evolution. Our modes of attack on the problem were different but our conclusions were much the same: possible, but rather fragile, under changing conditions. I came to appreciate the importance of this question much more fully after reading your 1962 book. I have no doubt that group selection has prevailed in many cases. I discussed it (much too briefly) in *Evolution and the Genetics of Populations*.[49]

This letter must have provided great satisfaction to Wynne-Edwards and spurred him to complete the manuscript the following year.

The next day, Wynne-Edwards wrote to Michael Wade at the University of Chicago, informing him of his forthcoming book and requesting permission to reproduce figures from Wade's 1977 paper in *Evolution*. Wynne-Edwards wrote enthusiastically of how Wade's experimental work provided "strong practical support" and "impressive elucidatory powers in accounting for the evolution of statistically (i.e. group) adapted genetic mechanisms, including the tight control of dispersal, and the separation of the two dimorphic sexes."[50] He had clearly identified Wade's work as a continuation and verification of Wright's model, importantly consistent with his own. Indeed, Wade concurred; in a review of *Evolution through Group Selection* in *Evolution,* Wade wrote: "Many of the studies of population structure in other organisms over the past decade have been presented in a conceptual vacuum, as though population structure per se, in the absence of any evolutionary perspective, had some intrinsic meaning. Wynne-Edwards must be acknowledged for his thorough presentation of the red grouse as a model for the population structure component of Wright's shifting balance theory."[51] Wade's assessment, however, was not echoed in the broader community.

Evolution through Group Selection: Magnum Opus II?

In a review written for the journal *Ethology,* Mark Ridley concluded, "What a strange man Wynne-Edwards is, to precipitate perhaps the most interesting controversy in evolutionary biology of the past quarter century, and then to write another book on the same subject, but which ignores the controversy completely."[52] Although Ridley gave Wynne-Edwards credit for his presentation of the grouse research that had been conducted for the previous thirty years at Culterty Field Station, as well as for his synopsis of the contemporary group selection work of David Sloan Wilson and Michael Wade, he pointed out that the book ultimately "contained no argument at all. It instead describes a point of view."[53] In another review in the *Journal of Genetics*, Mary Jane West-Eberhard concurred with Ridley's criticisms: "As a review of twenty years of research on the social behavior and ecology of the red grouse this book can be recommended*; as a review of the twenty years of thought on levels of selection it is most disappointing*. It will not satisfy readers who believe that a scientific argument is acceptable only if it deals perceptively and thoroughly with previous and contradictory alternative hypothesis."[54]

These criticisms were reiterated almost unanimously by the other reviewers of *Evolution through Group Selection.* West-Eberhard also made insightful criticisms regarding Wynne-Edwards's attempt to distinguish his group selection explanations of various social behaviors from William D. Hamilton's kin selection explanations of the same behaviors. She went on to point out that the distinction here was sometimes merely semantic and asserted that had Wynne-Edwards been willing, as were contemporary group selectionists David Sloan Wilson and Michel Wade, to view kin selection as a form of group selection or at least another type of selection that need not function counter to group selection, the biological community might have been more sympathetic. Unfortunately for Wynne-Edwards and his professional reputation, he was unable or unwilling to adapt his position on group selection to the many advances that had been made, and he essentially wrote himself out of the debate.

Notwithstanding Ridley's and West-Eberhard's supportive comments regarding the presentation of the grouse research that made up the seven central chapters of the book, a critical review by an unexpected author caused Wynne-Edwards great concern. Writing for the *Journal of Animal Ecology*, Adam Watson, who had been at the Culterty Field Station since its inception and had accompanied Wynne-Edwards on the 1953 expedition to Baffin Island, wrote a review that was to sour the relationship between the professor and his former student. The process began with a letter from Roy Taylor, editor of the *Journal of Animal Ecology*, on December 8, 1986. Taylor wrote to notify Wynne-Edwards of Watson's critical review. "I understand from Adam Watson that you have seen and accepted, perhaps not enthusiastically, his review of your book. In case you have any last minute comments I enclose a copy."[55]

This letter from Taylor, or more accurately, the copy of Watson's review, elicited a quick response from Wynne-Edwards. In prepublication correspondence with his editor at Blackwell Scientific Publications, Wynne-Edwards had already addressed a number of critical anonymous reviews. By this point he was well prepared for any number of contingencies. Nevertheless, the criticisms of his interpretation of the red grouse work, coming from his friend and colleague, were a painful blow.

On December 17 Wynne-Edwards sent a letter to Watson with a copy to Taylor. Most of it addressed, point-by-point, criticism in Watson's review, but it began with some general advice: "Reading [the review] again after an interval I am a bit upset at the hostility of its tone, and my first comment to you is that if my own hostile review of Frank Darling's 'Highlands and Islands' book many years ago is anything to go by, you might live, as I did, to wish you had not been so aggressive."[56]

Although the tone of Watson's review was not egregious, given the history between the two men even slight aggression would have been off-putting to Wynne-Edwards. In the letter he took pains to address each of Watson's criticisms and to point out where he had been misinterpreted or where Watson had missed the point entirely. In some passages the personal distress the review had caused came through clearly. On the second page Wynne-Edwards directly challenged Watson's claim regarding the establishment of the Culterty Field Station in 1956. In his review Watson rejected Wynne-Edwards's assertion that the field station was established to test his theory of group selection. Wynne-Edwards's comment also provides brief insight into his position on scientific disputes, or at least his position on this particular disagreement.

> On p. 3 you say, quite truly again, that I only once acknowledge that my interpretations are different from yours. The reason for this is that I believe no good comes of parading disagreements unnecessarily. Disagreement itself, however, is not a crime. Incidentally, who are you to say for what reason the grouse re-

search was started (lines 10–12)? In 1955–6 I desired above all else to find a way of testing my new hypothesis, and nothing would have diverted me from it. The approach from the Scottish Landowners came like a gift from heaven; but had it not offered the opportunity I wanted it would have been declined. The first sentence in chapter 7 is exactly correct.[57]

The sentence referred to states that the Culterty Field Station was established in 1956 "with the principal scientific aim of testing the *Animal Dispersion* hypothesis, or as much of it was amenable to test."[58] As Regius Professor in the Department of Zoology, Wynne-Edwards considered the establishment of the research field station something of a personal accomplishment. Furthermore, although he was perhaps willing to accept certain criticism of his science, he was decidedly not willing to entertain an alternative interpretation of his institutional achievement. He concluded: "I am well aware that the book still has many imperfections; but having spent so many years in search of facts, corroboration, and clarity of thought, I feel it reasonable to ask for rather more careful objective criticism. I am copying this to L. R. Taylor."[59]

The collection of Wynne-Edwards's papers contains only two versions of the review Watson wrote. According to a later letter to Taylor, Wynne-Edwards claimed Watson had sent him as many as five versions. This file includes one of the early drafts, although not the first, and the published version. Most of the revisions consisted of minor details and a stylistic adjustment that moderated the initially harsh tone. On December 23, Taylor wrote to notify Wynne-Edwards that he had spoken by phone with Watson and the review would be held back for revisions. He also acknowledged that he did not want to interfere in "what must be a very personal relationship." The last letters in this group of correspondence once again exhibit Wynne-Edwards's single-mindedness in the pursuit and in protecting his intellectual position. We also see a human side that has not been so apparent. In his short letter to Roy Taylor he wrote:

Thank you very much for the care you have taken to keep me informed about Adam's successive reviews of my book. This final one is a great improvement on any of his previous versions. He has spent hours & hours trying to put something together that is fair and balanced while presenting his own objections. I think I have actually seen five editions. He still makes statements that it would be easy to refute; but he has nevertheless leaned over backwards to be fair & I regard the discussions that the two of us have had together as having been very successful. Fortunately we are both temperamentally capable of cool and rational debate, though we each hold stubbornly to our irreconcilable views. Close friendships of 40 years can stand a bit of battering![60]

This short note, along with the letter to Watson two weeks later, shows that while Wynne-Edwards would always be an unyielding advocate of his theory, he would make every effort to keep personal and professional relationships above the fray. In his final note to Watson he wrote:

As he may have told you, Roy Taylor again sent me a copy of the final version of your review. This is just a brief note to say how much I appreciate the hours and hours you must have devoted to the job, first and last, and also the fairness of the criticism you have finally achieved. I shall be surprised if yours does not emerge in the end as the most penetrating and objective review the book receives. It is a pity we are going to have to wait so long to see it in print.[61]

This exchange between Taylor, Watson, and Wynne-Edwards was almost typical of the difficult reception that *Evolution through Group Selection* received. As I mentioned above, the prepublication reviewers were, on the whole, critical of the book and characterized it as more a work of advocacy than fresh scientific work. This was to be the legacy of his second book: Wynne-Edwards was essentially dismissed by many as an advocate of an outdated theory. The correspondence between Wynne-Edwards and Robert Campbell, his editor at Blackwell, shows that to a large extent Wynne-Edwards's acknowledgment of the debate that swirled around the theory of group selection was initiated by his editor, given the suggestions of manuscript reviewers. Campbell suggested in several letters that Wynne-Edwards might engage his contemporaries—both supporters and critics. Furthermore, early drafts of the manuscript mentioned the work of David Sloan Wilson as supportive of the theory of group selection, although not exactly the same theory as Wynne-Edwards's, but they missed altogether the work of Michael Wade. In a June 1984 letter to Campbell, Wynne-Edwards defended his not engaging with contemporary debates: "My policy has been to include enough of the vast literature to support my arguments, and enough to feel reasonably sure that my arguments have not been shot down already by others, *and let the rest go*. Other people's theories of, and objections to, group selection are not necessarily relevant to my particular thesis."[62]

Herein lies the problem. The first point to make is that most of the book is not aimed at dealing with contemporary criticisms of Wynne-Edwards's ideas; it simply documents further types of observations that make sense if group selection is a potent force in nature. Darwin's belief in evolution through natural selection was based on a consilience of inductions: myriad disparate facts suddenly made sense if evolution had occurred and was occurring through natural selection. Darwin was also aware that the contemporary theories of heredity and issues of the age of the earth were destructive to his theory. Nevertheless, Darwin stuck to his guns because so much about the natural world could be explained by his one unifying theory. As was pointed out in the *TLS* review at the beginning of chapter 6, Wynne-Edwards was in a very similar position. He saw a variety of behavioral patterns in the natural world that all made sense according to his theory of group selection.

In the course of this chapter it becomes remarkably clear that despite the debate continuing around him, Wynne-Edwards did not really engage. He remained resolute in his belief that the biological community was set against him and resisted prodding from his editors and his reviewers to address the new elements of the debate that had developed in the biological and the philosophical literature. It was clear by the 1980s that philosophers of biology had developed some useful distinctions and clarifications regarding group selection theory and the units of selection debate, and that these contributions had been acknowledged and incorporated by a new generation of biologists working on group selection theory. Wynne-Edwards wrote *Evolution through Group Selection* almost as if none of this had happened. I think this might have been the lowest point for Wynne-Edwards; nevertheless, he carried on.

CHAPTER EIGHT

The Death of Wynne-Edwards and the Life of an Idea

Mixed Messages

The resounding silence that greeted *Evolution through Group Selection* was nearly overwhelming for Wynne-Edwards. He had been retired from the University of Aberdeen for just over ten years, and this book was an attempt to reenter the academic arena and settle the debate over group selection once and for all. Another project may have led him to expect better for his second book; in 1985 Donald Dewsbury published *Leaders in the Study of Animal Behavior: Autobiographical Perspectives.*[1] Wynne-Edwards's autobiography was included along with those of such well-known ethologists as Konrad Lorenz, Niko Tinbergen, and Irenaus Eibl-Eibesfeldt. Perhaps this sanctioning would help persuade people to give his second book careful attention?

It was not to be; after the first round of reviews the book generated little professional interest and no public response whatever. Wynne-Edwards worked hard to generate interest in the book; he arranged for translations to be made and pushed journal editors to publish a précis similar to the précis of *Animal Dispersion* that *Scientific American* had published in 1965. His efforts went unrewarded. Much of his correspondence dealt with his disappointment over the lack of response, and the general silence convinced Wynne-Edwards that the community of evolutionary biologists had to ignore his book because they could not refute his theory. Writing to Elliston Perot Walker, a member of the Macmillan expedition he had not communicated with for some thirty years, Wynne-Edwards described the reception of his book a year after its publication. Walker had written to Wynne-Edwards asking for news about him, since he had missed the fifty-year reunion of the Macmillan expedition held in May 1987 at Bowdoin College in Maine. Wynne-Edwards had attended and exchanged memories and stories with fifteen other members of the expedition.

> As I have inserted on the last page of the offprint, my second book came out in April 1986. It revives the group selection controversy that most evolutionists had deluded themselves into thinking was dead and buried several years ago, and which they have been loud in disclaiming as heresy. Consequently they tend to regard its resurgence with resentment and hostility. In fact, there is still a deafening silence from the top-line leaders in the field. With one or two exceptions the task of reviewing it had been turned over to people with less reputation at stake. Some of the short notices that have appeared have commended its readability and lack of jargon, but without venturing to come off the fence as to its credibility. The only really ace reviewer, Lawrence Slobodkin of the Department of Ecology and Evolution at SUNY, Stony Brook, concluded, in *American Scientist*, albeit reluctantly, that its message cannot be dismissed and will have to be taken seriously.[2]

This characterization of the reception of the second book reveals a good deal about the author. Wynne-Edwards's optimism and resentment are readily apparent. The reviews described in the preceding chapter notwithstanding, he still presented his case as a strong one and, moreover, one that had yet to be successfully refuted. The note also indicates something about Wynne-Edwards's unawareness of his own shortcomings. By 1987 he had spent the better part of three decades talking past his critics, never engaging in the debate under the terms that had developed through the work of other group selectionists, but continuing to insist that his original formulation of the theory of group selection had never been successfully challenged.

Wynne-Edwards spent the late 1980s and early 1990s corresponding with the editors of journals, seeking one more opportunity to make his case for group selection. In 1991 he published an article in the *Ecologist* titled "Ecology Denies Neo-Darwinism." The article generated a response from two ecologists at the University of Michigan—Joel Heinen and Bobbi Low. Although this response was never published, it provides an excellent example of the continuing criticism Wynne-Edwards received. The authors of the critique characterized the article as "no less than a challenge to all modern biological theory."[3] They began their response with the assertion that "more than anyone else, Wynne-Edwards did much to cause evolutionary biologists to think critically about the units of selection with the publication of his well-written 1962 volume *Animal Dispersion in Relation to Social Behavior*. The problem is that he was profoundly wrong in 1962, and is profoundly wrong still."[4] In their eight-page response to Wynne-Edwards, Heinen and Low reiterated many of the most common criticisms of his later work, especially that he had failed to engage with the controversy generated by his first book. They argued that neo-Darwinian theory had been ultimately established and most clearly articulated in *The Selfish Gene*. The view of selection acting at the level of the gene, or in some cases the individual, was the cornerstone of contemporary evolutionary theory, and Wynne-Edwards was apparently alone in rejecting this. They also argued, contrary to his assertions, that natural populations are not stable. In another passage Heinen and Low cited Wynne-Edwards to illustrate his fundamental misunderstanding of neo-Darwinism. They quoted this passage from the 1991 article, referring to neo-Darwinism: "If that were the whole story, the traits one would expect to predominate would be ruthless pursuit of self-advantage and prolific fecundity."[5]

Heinen and Low argued that claims like this reflected Wynne-Edwards's inability to properly understand neo-Darwinian theory. In the margin of his copy of their response, Wynne-Edwards pointed out that the claim cited had been qualified in the article. The line reads as quoted, but the end of the sentence, not included by Heinen and Low, reveals a better understanding than they were willing to allow. The whole paragraph from which the partial sentence is cited presents a more sophisticated understanding of the neo-Darwinian theory than Heinen and Low suggested in their citation:

> Neo-Darwinian evolutionists, however, hold firmly to the belief that natural selection can operate only on individual organisms, and that all the adaptations and advances which evolution has witnessed must have arisen by that process alone. This in turn implies that the changes in gene frequencies from generation to generation must all, in the first instance, have increased the fitness of

> the individuals. If that were the whole story the traits one would expect to predominate would be a ruthless pursuit of self-advantage and prolific fecundity, *in so far as the latter allowed for more offspring to attain parenthood in their own generation.*[6]

This characterization of the neo-Darwinian paradigm does not appear particularly inaccurate. Heinen and Low's critique concluded with the implications for human ecology according to Wynne-Edwards and argued that this example provided the most damning evidence against his theory. In a rather simplistic tone, they argued that if humans were subject to group selection, "our populations would be stable, and we would be very willing to incur costs, such as higher taxes, so that the other group members would be better off. It is doubtful that the likes of Thatcher/Major, Reagan/Bush, Mulroney, Kohl, or Takeshita/Kaifu would have been elected, and it is doubtful that the glorious socialists of the Soviet Bloc and China would have turned out to be so corrupt."[7]

This argument shows a level of reasoning that is something less than rigorous, especially given that this is Heinen and Low's area of specialty. They are particularly interested in environmental resource management and the application of evolutionary theory to human behavior. It is for this reason, they argue, that they felt compelled to respond so harshly to Wynne-Edwards's article. This exchange provides another nice example of Wynne-Edwards's seeming inability to incorporate the advances in evolutionary theory while also presenting Heinen and Low as the most recent in a long line of critics who had demonized a caricature of him.

The note that Patrick McCully, coeditor of the *Ecologist*, included with Heinen and Low's critique invited Wynne-Edwards to reply. His response to McCully opened with the following salvo.

> I have had a good look at the comment by Joel Heinen and Bobbi Low on my group selection article. It is muddleheaded and full of errors and misquotes. They have not made any serious attempt to understand or fathom the article and on several occasions directly ignore or even reverse particular points that are made in it. In spite of their frequent insistence that distinguished biologists have ideas that are different from mine, none of these paragons named have yet "refuted Wynne-Edwards" even in their reviews of the book in 1986. . . . If you think you can get something publishable out of them I should like to be allowed to reply: but it would be an unenjoyable task.[8]

Ultimately McCully decided not to publish Heinen and Low's piece and instead printed two much shorter reviews. The first was incredibly enthusiastic and exclaimed that "Professor Wynne-Edwards has, unwittingly perhaps, done for the 21st century what Karl Marx sought to do for the 19th, he has provided a manifesto for the political needs of the time."[9] The second was unequivocally critical and invoked the selective value of each genotype and William of Occam's razor to carry the argument.[10] Given this even-handed treatment of his work by the editor and Wynne-Edwards's conviction that group selection had never been successfully refuted, his continued campaign to publish was completely predictable.

The article published in the *Ecologist* was not the one Wynne-Edwards had been hoping for. He had a much longer and better-developed treatment of the theory in mind and spent the next couple of years pushing for a venue. He pursued this course despite cataracts in both eyes and a retinal hemorrhage in the right. This situation caused a slowdown in the eighty-four-year-old's work schedule, but as he recounted in a letter to a former colleague from Aberdeen, he was not easily deterred:

> You know I have been an advocate of group selection as a powerful evolutionary force, for more than 30 years. I published a second book in 1986 entitled *Evolution Through Group Selection*. It did not succeed in its mission and was not a success, as *Animal Dispersion* had been. Since then I have found the biology "establishment" against me, and editors refusing to publish the three or four papers I wrote: until late in 1992 *The Journal of Theoretical Biology* was persuaded to invite me to submit a review. I spent most of 1993 preparing a mini-review, of about 12,000 words (22 pages). (My blindness came on before I had finished it!) But it was eventually submitted, accepted, and published in May 1993. It is attracting much attention, and I am in high hopes that my explanation of how group selection works will be accepted this time. No one has offered to challenge it so far, and I feel almost certain that I am going to have the last word! Certainly my 20 years of "retirement" have never had a dull moment yet.[11]

The story of the initial rejection and ultimate publication of the mini-review in the *Journal of Theoretical Biology* is corroborated by Wynne-Edwards's collection of correspondence with editors and friends in the late 1980s and early 1990s. In a letter to the editor of *Nature,* he pleaded his case from two perspectives. First, he argued that since the publication of his second book and its poor reception, the subject of group selection had been effectively stifled. In one passage from his letter he claimed,

"Virtually all the leaders of opinion had been confidently committed to the alternative [to group selection] 'individual selection only' Neo-Darwinist dogma, to the extent that any support for group selection had come to be denounced as heresy. I believe that some of them at least are embarrassed now being confronted with the entirely unforeseen possibility of being proved wrong."[12]

His second point was that he was too old at this point to travel to conferences and present papers, and that left the published paper as the only mode through which he could effect a debate. He wrote that "something must be done to stir things up."[13]

Wynne-Edwards's initial offer of his completed manuscript prompted John Maddox, the editor of *Nature*, to agree to read it and respond. The packet of correspondence between Wynne-Edwards and the editors at *Nature* is particularly interesting because the copies Wynne-Edwards had in his file appear to have come from the editorial offices at *Nature*. His first inquiry was written on March 16, 1989, and the copy in the collection includes the handwritten and underlined notation, NO across the top. On the copy of the follow-up letter written June 14, 1989, in which Wynne-Edwards requested an update, there is a handwritten message at the bottom labeled URGENT. The message reads as follows:

> HG → JM [John Maddox] 15/6/89
> This letter refers to the one below discussed in N + V [News and Views] meeting 21/6/89. It was decided that you would send a tactful "no" to the author. Did you send one? If not, please do so. Thanks.[14]

This memo was followed by the letter from John Maddox in which he wrote to Wynne-Edwards on June 30 that since the manuscript was already written, Maddox would read it himself and give his opinion on whether it was publishable. He concluded his letter, contrary to the instructions in the memo above: "But since you have a manuscript ready, why not send it? I would read it personally and if I thought there was a chance of persuading some of our referees to give it a good reading I would try it on them."[15] Wynne-Edwards responded three days later and mailed the manuscript of "The Evolution of Cooperative Behavior." It was more than three months before he got final word. In a letter dated October 23, Maddox apologized for all the delays and wrote: "Sadly, despite my enjoyment of it I must tell you straight away that we cannot publish it."[16] At the end of the paragraph he offered what he hoped was a reasonable explanation. Essentially, he explained that Wynne-Edwards's argument for group selection in relation to population density was sound logically and self-consistent, but it did not offer any reason for biologists committed to individual selection to change their point of view. It was the same critique offered so successfully by George C. Williams thirty-three years earlier. All the phenomena explained by Wynne-Edwards's account of group selection could be as effectively explained at the individual level; the principle of parsimony and an overall trend toward reductionism in biology led to a rejection of the theory.

Despite this rejection, Wynne-Edwards was characteristically undaunted; within ten days he had sent the manuscript to Andrew Sugden, the editor at *Trends in Ecology and Evolution*. The response here was unusual, because Sugden wrote that though he could not publish the article, his reasons had nothing to do with the scientific merits or interest of the work. Rather, he pointed out that the manuscript was too long and, contrary to editorial policy at their journal, concentrated too much on the author's own work. He concluded his rejection with an invitation to write a shorter piece that would provide a general review of recent thinking on the group selection question. This of course was exactly the opposite of Wynne-Edwards's intent. He had no desire to engage in the contemporary debate per se; rather, he was doggedly seeking a venue for reintroducing his own theory.

Shortly thereafter, Wynne-Edwards once again sent the manuscript out for consideration, this time, to Felicity Huntingford, the editor of *Animal Behaviour*. The result was the same, but in this instance Huntingford included the referees' reports. Here again the argument was essentially the one given by John Maddox at *Nature*. Nevertheless, one referee offered some insightful comments on the reaction to Wynne-Edwards's most recent work. The comments illuminate Wynne-Edwards's status in the biological community and the difficulty members of the community had in responding to his perpetual advocacy of the theory of group selection. In this case the referee is willing to give him credit not for developing the theory, but for pushing theoreticians to provide "proper" explanations for sociobehavioral phenomena.

> I fear that the silence [that greeted the second book] is best explained not by the fact that he is being ignored by the evolutionists but that they disagree with him but would rather not say so to one whom we all regard with great respect and affection.

FIGURE 15. V. C. Wynne-Edwards's Royal Society Fellowship portrait. *Biographical Memoirs of Fellows of the Royal Society* 44 (1988). Copyright © Godfrey Argent Studio.

> Wynne-Edwards is a great man, whose influence in the field has been enormous. Before the publication of his first book, and for a period thereafter, many evolutionists (led by David Lack) fervently believed that natural selection only acted at the level of the individual. "Animal Dispersion" Changed all that. It documented the fact that indeed "cooperation is . . . rife in the animal world" as he states here. It was his great contribution to point this out and a challenge to the neo-Darwinists to explain it.[17]

This was to be the final rejection of this manuscript. In July 1991 Wynne-Edwards received a letter from his old friend and colleague J. Z. Young. As an editor of the *Journal of Theoretical Biology,* Young implored Wynne-Edwards to submit a mini-review on the topic of group selection: "Would you do one for us? It would reach a huge audience of scientists of all sorts. . . . You could write up the theme explaining in rather [illegible word] detail the general evidence (beyond birds) and the weakness of the criticisms. Do please try. . . . Surely it would be good for the subject—Which is what matters."[18]

Of course Wynne-Edwards replied that he "would spare no effort to take advantage of the extremely kind offer." But he was cautious regarding the ultimate publication of any review he would write.

> I have learnt to expect that my opponents would nip the project in the bud. I am up against a very powerful lobby and have in fact had rejections from the editors of Scientific American, Nature and Animal Behaviour over the last three years. You are of course a very powerful ally, but the establishment (Oxford Zoology Department plus John Maynard Smith) will probably argue that evolution is not your field of expertise. Nevertheless, do try.[19]

Wynne-Edwards's last article, "A Rationale for Group Selection," was published in the *Journal of Theoretical Biology* in 1993. He was eighty-six years old. The article took up the first twenty-two pages of the journal and reinvigorated Wynne-Edwards, who had recovered his sight after lens-replacement surgery.

The 1993 article did not stimulate a revolution in evolutionary theory, but the response was generally positive. While Wynne-Edwards made use of Michael Wade's experimental results in his argument, he did not cite the work of David Sloan Wilson. This had more to do with the mathematical sophistication of Wilson's models than any real theoretical disagreement Wynne-Edwards may have had with him. Both Wynne-Edwards's son Hugh, a geologist, and his granddaughter Katherine, a behavioral endocrinologist, felt that the increasing importance of mathematical modeling in the study of animal behavior and population were a frustrating factor for Wynne-Edwards.[20] His training as a naturalist left him ill-prepared to assimilate these new and complicated techniques so late in life.

The End of the Crusade

Wynne-Edwards spent his final years of retirement in Scotland with his wife and lifelong companion Jeannie. According to the family he kept up his habits of watching birds and maintaining his bird logs until his death on January 5, 1997, at age ninety-one. Wynne-Edwards was elected a fellow of the Royal Society in 1970 and was awarded the Commander of the British Empire in 1973 for his long and distinguished service to science. His eulogizers lauded him for his erudition and his eagerness to encourage the intellectual development of others. Even those he disagreed with noted that he never interfered in others' scientific work; his approach was to choose good people and ensure them the funding and intellectual freedom to carry out their work.[21] That is not to say that Wynne-Edwards did not continue to seek converts to his theory until the very end. In a letter to Robert May, Royal Society Professor in the department of zoology at Oxford, Wynne-Edwards made a final attempt to pass the mantle. After a brief introductory paragraph, he wrote:

> There is of course a massive objection to accepting [group selection theory] from the majority of biologists who have been brought up to regard it as a heresy, although on dogmatic not experimental grounds. They seem unlikely to change their minds in a hurry unless they are given the true message loud and clear by authorities whose competence they trust. I am 87 years old and can play no further part. In the last eight years I have grown used to having articles rejected by editors who, being unqualified to make independent judgements, just assume I am a heretic.
>
> You and your Oxford colleagues constitute a centre of outstanding excellence in this field of knowledge. Though it might involve some back-tracking, you could if you saw fit, take over the conversion task as your own, with lasting benefit to scholarship at large, and great delight from this Oxonian son. But perhaps I presume too much. Yours Sincerely.[22]

In his reply, May thanked Wynne-Edwards for the letter but explained that he had enough contentious issues pertaining to his own interests and was reluctant to embrace any new crusades.

It is fascinating to see how this correspondence included religious imagery. Wynne-Edwards used the terms dogma, heresy, and conversion. May replied in kind, with crusade. Much of the current debate over group selection acknowledges this aspect of the controversy at least implicitly. To many it is a debate over where virtue lies—with the group or with the individual. It is an attempt to locate and explain the evolution of morality. This is part of the reason the debate persists, and maybe the main reason it will continue. There are models that show the effectiveness of group selection as an evolutionary mechanism, and supporting experimental results. The field evidence is less readily collected. Groups are difficult to study in the field even for a single season, let alone long term. Darwin recognized this, as did every subsequent naturalist. However, before the advent of sophisticated population modeling, this did not deter them from global theorizing. As Adam Watson and David Jenkins pointed out in their obituary of Wynne-Edwards, "He openly believed there was great value in using inductive reasoning for surveying natural phenomena, looking for evidence that confirmed ideas rather than refuting them and producing general explanations of the widest possible relevance."[23]

Wynne-Edwards maintained a commitment to the theory of group selection for nearly forty years despite an unrelenting current of evolutionary theory flowing in a more reductionist direction. Wynne-Edwards and his advocacy of group selection theory constitute an interesting case study. His persistence is both his strongest asset and his fatal flaw.

Vital Contribution or Garden Path?

The debate over group selection continues. Currently, however, the hierarchical nature of evolutionary theory is more accepted, and the role of group selection is grudgingly allowed. But what does this case tell us about the nature of theories, their development and their histories? What is the role of the individual scientist or groups of scientists in establishing or rejecting these theories? In the course of my research, these are some of the questions I hope I have addressed. As historian of biology Frederick Churchill once wrote:

> It is not enough for the historian to track down specific theories or metaphysical digressions; he must draw forth from his historical quarry a confession about his feelings toward the scientific estate. To observe, to experiment, to employ sister disciplines as implements of discovery, or to insist on one approach and to exclude another, tells much more than the empirical contents of a given piece of research. Such a confession tells what the scientist believes to be the boundaries of scientific knowledge.[24]

What have we gleaned from examining the development of the theory of group selection through the lens of Wynne-Edwards's career? His commitment to a hierarchical approach to evolutionary theory was the result of a number of contributing factors. Furthermore, his inability to accommodate the implication that arose from developing mathematical models and philosophical critiques of those models can also be understood in terms of these same influences.

The importance of Wynne-Edwards's work is not disputed. Indeed, in a 1989 article in the *Quarterly Review of Biology*, Robert McIntosh reviewed the citation classics in ecology and found that during 1962 to 1980 *Animal Dispersion* had been cited 600 times, for an average of 35.3 citations a year. Although two other works had higher averages, Wynne-Edwards's had the highest total number of citations of the eighty works included.[25] Of course the analysis did not include the way the citation was used. McIntosh reminded his readers, "It is important to note if, and what aspect of, a cited work was mentioned in the citing article and whether it was supported, negated, or conceivably even used as a horrible example."[26] Regarding *Animal Dispersion*, it is safe to assume, given all we have seen thus far, that a significant portion of the citations must have been negative. But this does not necessarily detract from the work's importance. Many thought Wynne-Edwards had provided a very important theoretical stalking horse for evolutionary theorists. The preceding generations' dependence on unexamined claims of behavior that could be explained in terms of the good of the species was finally tested. In the received view, these claims not only were put to the test but were ultimately laid to rest. Obviously this is not particularly accurate, and here is another lesson we can derive from the history of group selection theory.

Wynne-Edwards's advocacy appears to have had a very deep and not necessarily consistent effect on the development of group selection theory. It clearly stimulated a negative response from those committed to neo-Darwinian individual-level selection. But Wynne-Edwards influenced some portion of the biological community to continue to examine the possibility that selection did indeed act on the group level and that individual-level explanations were not always sufficient. Both David Sloan Wilson and Michael Wade have acknowledged the influence of Wynne-Edwards on the early development of their thinking.[27] In both these cases, however, his dogged commitment to his own formulation of group selection theory limited his influence and his contribution to the debate as it developed through the 1970s and 1980s.

Theory Ladenness and Observation

In the 1950s, Norwood Russell Hanson brought the notion of the "theory ladenness of the observer" to prominence in the philosophy of science. In Hanson's scenario, Johannes Kepler and Tycho Brahe stood together at dawn. Kepler regarded the sun as fixed; Brahe followed Aristotle and Ptolemy, and for him the earth was fixed and all other celestial bodies moved around it. This raised the question, Did Kepler and Brahe see the same thing in the east at dawn? This question continues to generate discussion, at least in introductory philosophy of science classes, and I think everyone agrees with Hanson's summation: "There is more to seeing than meets the eyeball."[28]

In the course of this story I have presented the evolutionary theories of two field naturalists: Prince Petr Kropotkin and Vero Copner Wynne-Edwards. Both these men developed theories that emphasized mechanisms *other* than individual-level natural selection to explain aspects of the natural world that were ill-suited to explanation by standard Darwinian theory. The theories these naturalists developed are noteworthy because both Kropotkin and Wynne-Edwards were theoretically prepped by exposure to the ideas in the *Origin* to see nature in a particular way; and yet *neither* of them saw what Darwin saw when he looked at nature.

A number of factors are at work here. First is the ecological region where each of these men conducted his research. Second is the state of biology as a discipline; and third is the broader sociopolitical/ideological context. I believe the prime motivator in the development of both their theories was their experience in the field. Both of these men worked in Arctic environs and experienced nature in a significantly different way than did either Darwin or his codiscoverer Alfred Russel Wallace. Nevertheless, the other factors clearly contributed.

Kropotkin's theory was initially developed as a response to Thomas Huxley's gladiatorial view of nature, as presented in his 1888 article "The Struggle for Existence in Human Society."[29] Over the next twenty years Kropotkin's main foe became the German biologist August Weismann. His theoretical target, however, remained consistent—that is, biologists' overemphasis on the role of natural selection in evolution.

Wynne-Edwards became intrigued by behavior among fulmars that struck him as inexplicable in Darwinian terms. In response to David Lack's 1954 book *Natural Regulation of Animal Numbers*, Wynne-Edwards developed his theory of group selection. This theory, consistent with Kropotkin's, provided an explanation of a wide range of animal behavior that was not dependent on neo-Darwinian ideas of natural selection and intraspecific competition.

And so we find ourselves back to Hanson. This time, however, we see two pairs of *naturalists* facing the east at dawn; the neo-Darwinians, Thomas Huxley and David Lack, and the pluralists, Petr Kropotkin and

V. C. Wynne-Edwards. Unlike Brahe and Kepler, their attention is captured not by the position of the sun, but by the colonies of birds on the adjacent cliffs. Again the question arises: Do they see the same thing? As I hope has become apparent in the course of this story, the answer must be that they *do not.* Where Huxley and Lack see an individual struggle (of each against all) to procure resources to provide for themselves, Kropotkin and Wynne-Edwards are witness to a cooperative struggle by social groups against the environment. Where Huxley and Lack see harsh elements increasing the intraspecific level of competition, Kropotkin and Wynne-Edwards observe a state of depleted resources that reduces the fitness, and threatens the survival, of the entire group.

Nevertheless, despite the clear differences between the pairs, both lay claim to the Darwinian heritage. Kropotkin's Siberian experience, the Russian naturalist tradition, and even his anarchist politics clearly influenced his non-Malthusian view of nature. These elements, compounded by the state of Darwinian theory in the early twentieth century, provided ample support and theoretical space for developing his theory.

The case is not so clear-cut for Wynne-Edwards: a product of Oxford and a student of Julian Huxley and Charles Elton; witness to the modern synthesis; and a member of the British middle class. Wynne-Edwards's theory presents something of a challenge. Essentially, I think his field experience led to the theory of group selection. The initial influence of the modern synthesis and an emphasis on population thinking may also have contributed. The hardening of the synthesis around the mechanism of natural selection acting at the level of the individual, however, created a hostile environment for the long-term survival of his theory.

The analysis of these two theories should be important to historians of biology for two reasons. First, because they illuminate the broader context of the development of Darwin's theory. Second, because the debate over the validity of group selection and its role in the evolutionary process continues. Historians have an opportunity here to contribute to ongoing discussions of the hierarchical nature of evolutionary theory. Finally, this history can be useful to students of biology. Many of the common misunderstandings of evolution are a direct result of the exclusive application of natural selection at the level of the individual.

In conclusion, Darwin's own explanation of the concept of struggle from the *Origin* is important here: "I use this term in a large and metaphorical sense including dependence of one being on another, and including (which is more important) not only the life of the individual, but the success in leaving progeny. Two canine animals, in a time of dearth, may truly be said to struggle with each other over which shall get food and live. But a plant on the edge of a desert is said to struggle for life against the drought."[30]

Hanson was right; there is more to seeing than meets the eyeball.

In a recent philosophical analysis of the influence of ideology in the group selection debate, philosopher Ayelet Shavit examines the ways that different images of the notion of group, with their social and political values, have shaped this debate historically and continue to inhibit the interpretation and evaluation of various empirical experiments and observations. She argues that the ideologically influenced image of the group as either "a loose aggregate that prioritizes the individual [or] a tight superorganism that undermines the independence of the individual" held by a particular researcher "projects on the way tests of group selection are described and assessed."[31] She identifies Hamilton, Williams, Maynard Smith, and Dawkins as viewing groups as extremely tight superorganisms. In the post–World War II era this view was associated with fascist ideologies, while individual selection was associated with democracy and freedom. On the other hand, the aggregate view of groups, held by Warder Clyde Allee, V. C. Wynne-Edwards, and later David Sloan Wilson and Michael Wade, "require[s] individual restriction, though such restriction is based upon contracts among free willing individuals."[32] She concludes:

> Leading figures of the group selection debate respond to echoes of past historical conflicts by polarizing the meaning of "group" into an extreme image connected to a fixed set of values. However, various gradations and combinations of superorganism and aggregate images are necessary for an accurate description of cooperative interactions. Adherence to a loose aggregate image by supporters of group selection, and to a tight superorganism image by opponents of group selection, makes empirical tests that successfully distinguish between competing [selection] processes less important.[33]

Indeed, it is apparent from the account presented in the preceding chapters that these sometimes explicit but more often tacit ideological commitments have been a significant force in the course of this debate. Unfortunately, the ambiguity regarding the proper characterization of groups persists in some of the most recent disputes regarding the relative significance of group selection in evolution.[34]

The Process of Science

In his 1988 book *Science as a Process*, David Hull takes a selectionist approach to the process of science. In his analysis of the development of the field of systematics, Hull identifies important roles for history in the understanding of science:

> The explanatory force in historical explanations comes from the coherence and continuity of the historical entities whose development they chronicle. In historical explanations either an entity is shown to be part of a more inclusive historical entity or else the course of the entity is described. In the first instance, an entity is placed in its historical context—a star in its galaxy, a fossil in its lineage, a scientist in his research group, or a term-token in its research program. In the second instance, an historical entity is traced through time. Historical narratives chronicle the development of historical entities through time. Although which entities count as entities is theory dependent, one need not always have a highly articulated theory to distinguish historical entities and follow them through time. Hence, even though historical explanations do not involve derivations from laws of nature, they are not theory independent either.[35]

Hull's characterization of history here is apt. I have attempted to treat two historical entities—V. C. Wynne-Edwards and his theory of group selection—in both of the ways described above.

The treatment of the individual was perhaps more straightforward in the sense that an individual is unproblematic in terms of its beginning and end as well as its location in space and time. The history of Wynne-Edwards began on July 4, 1906, and ended on January 5, 1997. I think the story of his professional life provides an important vantage point from which to analyze the development of the theory of group selection.

As we have seen in the course of this narrative, the history of the idea of group selection predates Wynne-Edwards and has outlasted him as well. Although his formulation of group selection has been surpassed, his contribution to the development the hierarchical understanding of evolutionary theory should not be undervalued. The vague assertions of adaptations for the good of the species at the turn of the century have become much more concise assessments of group fitness and group-level adaptations evinced in sophisticated mathematical models and impressive experimental results. Wynne-Edwards's advocacy of the importance of understanding the role of social behavior in the evolution of population structure served as a catalyst for further study of this important relationship.

Although some might argue that Wynne-Edwards's unwavering commitment to his idea of group selection did more harm than good to the development of the theory, I think my analysis shows that this is incorrect. He may have diminished his own professional status as a result of his steadfast insistence on the importance of group selection, but he also helped to create the theoretical space where subsequent researchers could develop more careful analyses. His continued emphasis on the importance of understanding biological phenomena occurring above the level of the individual has been fundamental to the development of ecology and behavioral biology.

Epilogue

In the decade since Wynne-Edwards's death there has been a continuing interest in the role of group selection in evolutionary history. Indeed, some of the most ardent critics of the theory have come to acknowledge that group selection has likely been essential in the evolution of sex and the evolution of social behavior.[36] Evolutionary biologists like Michael Wade and David Sloan Wilson, who have been arguing for the importance of group selection for decades, have been joined by a growing number of researchers in a variety of fields who have developed group selection models and applied them in primatology and anthropology, behavioral economics, human behavior, and population ecology.[37] Indeed, it is becoming increasingly common in the literature to invoke group selection with respect to human evolution. Some recent papers have even revived and reexamined Wynne-Edwards's specific hypothesis regarding population structure, social communication, and reproductive restraint.[38] These studies, which combine computer simulations and experimental treatments of microbial populations, were not technically feasible ten years ago and have caused a broad range of evolutionary biologists to reexamine the significance of group selection. These particular studies have been accompanied by more general treatments of evolutionary theory that have also emphasized the significance of group selection.[39] In his extended plea for a revision of evolutionary theory toward a more hierarchical approach, Stephen Jay Gould was explicit regarding group selection, and Wynne-Edwards in particular: "The ingenuity of Wynne-Edwards' theory lies largely in the range of behavioral phenomena that he interprets as devices evolved by group selection for limitation in population size."[40] This characterization of Wynne-Edwards's contribution is consistent with what the philosopher Samir Okasha has identified as the most significant and productive development in the recent work on the levels of selection question. He argues that connecting the levels of selection question to the work on major transitions in evolution may bring about much needed clarity and progress. The move from a synchronic approach (that takes the existence of levels for granted) to a diachronic approach (that provides an evolutionary account for the existence of the higher levels) requires a much more careful analysis of the levels at which natural selection is acting. He concludes his account by citing developmental biologist C. H. Waddington:

> Whatever the future developments in the field look like, it is likely that multi-level selection will remain crucial for theorizing about evolutionary transitions. Given that multi-level selection theory, in its current form, is the product of many decades of work on the levels-of-selection problem in biology, it follows that this work has left an important intellectual legacy. C. H. Waddington famously dismissed the levels-of-selection debate of the 1960's as a "rather foolish controversy," an opinion in which he was not alone. The centrality of the levels-of-selection issue in the contemporary literature on evolutionary transitions shows that Waddington's opinion was seriously mistaken.[41]

I couldn't agree more.

Notes

Introduction

1. David Sloan Wilson, "The Group Selection Controversy: History and Current Status," *Annual Review of Ecology and Systematics* 14 (1983): 159–187. See also Mark E. Borrello, "The Rise, Fall and Resurrection of Group Selection," *Endeavour* 29 (2005): 43–47.

2. Michael Ruse, "Charles Darwin and Group Selection," *Annals of Science* 37 (1980): 615–630.

3. Charles Darwin, *On the Origin of Species* (Cambridge, MA: Harvard University Press, 1859), and Charles Darwin, *The Descent of Man, and Selection in relation to Sex,* vols. 1 and 2 (Princeton, NJ: Princeton University Press, 1871).

4. Jean Gayon, *Darwinism's Struggle for Survival: Heredity and the Hypothesis of Natural Selection,* trans. Matthew Cobb (Cambridge: Cambridge University Press, 1998), 68.

5. Though Gayon has recently reconsidered his strong position on this point. Personal communication, July 2009.

6. Theodosius Dobzhansky, *Genetics and the Origin of Species,* ed. Leslie C. Dunn (New York: Columbia University Press, 1937).

7. Ibid., 127.

8. V. C. Wynne-Edwards, "The Nature of Subspecies," *Scottish Naturalist* 60 (1948): 195–208, 195–196.

9. It is important to point out that the "physiology of populations" was explicitly the research program of University of Chicago ecology professor Thomas Park. This is significant because Park is seen as a father of experimental ecology, and eight of his papers published by 1938 were titled "Studies in Population Physiology." I am indebted to an anonymous reviewer for this point.

10. Ibid., 120.

11. Lee Alan Dugatkin, *The Altruism Equation: Seven Scientists Search for the Origins of Goodness* (Princeton, NJ: Princeton University Press, 2006), 99.

Chapter One

1. Frederick Churchill, "The Weismann-Spencer Controversy over the Inheritance of Acquired Characters," paper presented at the Fifteenth International Congress of the History of Science, Edinburgh, August 10–15, 1977.

2. Michael Ruse, "Charles Darwin and Group Selection," *Annals of Science* 37 (1980): 615–630.

3. Ibid., 617.

4. See, for a recent example, Piter Bijma and Michael Wade, "The Joint Effects of Kin, Multilevel Selection and Indirect Genetic Effects on Response to Genetic Selection," *Journal of Evolutionary Biology* 21 (2008): 1175–1188.

5. Charles Darwin, *On the Origin of Species* (Cambridge, MA: Harvard University Press, 1859); emphasis added.

6. Ibid., 236; emphasis added.

7. Ibid., 241.

8. Thomas Dixon, *The Invention of Altruism: The Making of Moral Meanings in Victorian Britain* (Oxford: Oxford University Press, 2008), 145.

9. Ibid., 147. Dixon essentially argues that historians have gotten Darwin's focus wrong with respect to the social insects. In *The Ant and the Peacock: Altruism and Sexual Selection from Darwin to Today* (Cambridge: Cambridge University Press, 1991), Helena Cronin argues that the problem of altruism was the focus of Darwin's work on the social insects. Dixon makes a compelling case for the idea that Darwin was primarily concerned with the physical traits of the various sterile castes (as described in the preceding passage) and only secondarily interested in "closing the gap between humanity and the rest of the animal world by showing, first, that the moral qualities of sympathy and love were present in the lower animals as well as in man and, secondly, that a plausible evolutionary account could be constructed of how those social instincts, in the case of man, could have risen to such a pitch as to produce those intense feelings of moral duty and obligation with which he and his Victorian readers were so familiar" (Dixon, *Invention of Altruism,* 146).

10. Darwin, *Origin of Species* (1859), 238.

11. Ibid., 203.

12. Charles Darwin, *The Descent of Man, and Selection in relation to Sex* (1871; Princeton, NJ: Princeton University Press, 1981), 103.

13. Ibid., 155.

14. Ibid., 162.

15. See Peter J. Bowler, *The Eclipse of Darwinism: Anti-Darwinian Evolution Theories in the Decades around 1900* (Baltimore, MD: Johns Hopkins University Press, 1983); Peter J. Bowler, *The Non-Darwinian Revolution: Reinterpreting a Historical Myth* (Baltimore, MD: Johns Hopkins University Press, 1988); Richard Hofstadter, *Social Darwinism in American Thought* (Boston: Beacon Press, 1955). On progress see Matthew Nitecki, ed., *Evolutionary Progress* (Chicago: University

of Chicago Press, 1988). On Malthus see Michael Mogie, "Malthus and Darwin: World Views Apart," *Evolution* 50 (1996): 2086–2088; Daniel Todes, "Darwin's Malthusian Metaphor and Russian Evolutionary Thought, 1859–1917," *Isis* 78 (1987): 537–551; and Daniel Todes, *Darwin without Malthus: The Struggle for Existence in Russian Evolutionary Thought* (New York: Oxford University Press, 1989). See also Robert J. Richards, *The Meaning of Evolution: The Morphological Construction and Ideological Reconstruction of Darwin's Theory* (Chicago; University of Chicago Press, 1992).

16. Philip Pauly, *Controlling Life: Jacques Loeb and the Engineering Ideal in Biology* (New York: Oxford University Press, 1987).

17. William B. Provine, *The Origins of Theoretical Population Genetics* (Chicago: University of Chicago Press, 1971).

18. August Weismann, *The Evolution Theory* (London: Edward Arnold, 1904).

19. Ibid., 89.

20. Michael J. Wade, "The Changes in Group Selected Traits When Group Selection Is Relaxed," *Evolution* 38 (1984): 1039–1046.

21. Weismann, *Evolution Theory,* 98.

22. David Starr Jordan and Vernon Lyman Kellogg, *Evolution and Animal Life: An Elementary Discussion of Facts, Processes, Laws and Theories relating to the Life and Evolution of Animals* (New York: D. Appleton, 1907).

23. Ibid., 387.

24. Ibid., 397.

25. Ibid.

26. Vernon L. Kellogg, *Darwinism To-day: A Discussion of Present-Day Criticism of the Darwinian Selection Theories, Together with a Brief Account of the Principal Other Proposed Auxiliary and Alternative Theories of Species-Forming* (New York: Henry Holt, 1908).

27. See Robert E. Kohler, *From Medical Chemistry to Biochemistry: The Making of a Biomedical Discipline* (Cambridge: Cambridge University Press, 1982).

28. Thomas Hunt Morgan, *Evolution and Adaptation* (New York: Macmillan, 1903), 350–352.

29. Thomas Hunt Morgan, *The Scientific Basis of Evolution* (New York: W. W. Norton, 1932). For detailed analysis of the development of Morgan's thought, see also Garland E. Allen, "Thomas Hunt Morgan and the Problem of Natural Selection," *Journal of the History of Biology* 1 (1968): 113–140, and Garland E. Allen, *Thomas Hunt Morgan: The Man and His Science* (Princeton, NJ: Princeton University Press, 1978); see also Robert E. Kohler, *Lords of the Fly:* Drosophila *Genetics and the Experimental Life* (Chicago: University of Chicago Press, 1994).

30. Kellogg, *Darwinism Today,* 83.

31. James Arthur Thomson, *Concerning Evolution* (New Haven, CT: Yale University Press, 1925).

32. Ibid., 120.

33. Ibid., 141.

34. For a contemporary discussion of nested selective sieves see Elliott Sober, *The Nature of Selection: Evolutionary Theory in Philosophical Focus* (Chicago: University of Chicago Press, 1984), 97–102.

Chapter Two

1. Edward D. Cope, "Heredity in the Social Colonies of the Hymenoptera," *Proceedings of the Academy of Natural Sciences of Philadelphia,* 1893, 437.

2. Ibid.

3. Ibid., 438.

4. William Ball, "Neuter Insects and Lamarckism," *Natural Science,* February 1894, 97.

5. Joseph T. Cunningham, "Neuter Insects and Darwinism," *Natural Science,* April 1894, 289.

6. Ibid., 288.

7. Ball, "Neuter Insects and Lamarckism," 97.

8. Charles V. Riley, "On Social Insects and Evolution," *Report of the British Association for the Advancement of Science,* 1894, 689–691.

9. Ibid., 691.

10. Ibid.

11. Charles V. Riley, "Social Insects from Psychical and Evolutional Points of View," *Proceedings of the Biological Society of Washington* 9 (April 1894): 53.

12. Ibid., 60.

13. For an excellent treatment of these issues, see Robert J. Richards, *Darwin and the Emergence of Evolutionary Theories of Mind and Behavior* (Chicago: University of Chicago Press, 1987). See also Thomas Dixon, *The Invention of Altruism: Making Moral Meanings in Victorian Britain* (Oxford: Oxford University Press, 2008); Charlotte Sleigh, *Six Legs Better: A Cultural History of Myrmecology* (Baltimore, MD: Johns Hopkins University Press, 2007); and Lorraine Daston and Fernando Vidal, eds., *The Moral Authority of Nature* (Chicago: University of Chicago Press, 2004).

14. William M. Wheeler, "The Ant-Colony as an Organism," *Journal of Morphology* 22 (1911): 325.

15. For a nice analysis of the superorganism metaphor, see Sandra D. Mitchell, "The Superorganism Metaphor: Then and Now," in *Biology as Society, Society as Biology: Metaphors,* ed. Sabine Massen, Everett Mendelsohn, and Peter Weingart (Dordrecht: Kluwer Academic Publishers, 1994), 231–247. For a recent reintroduction of the superorganism concept, see Bert Hölldobler and Edward O. Wilson, *The Superorganism: The Beauty, Elegance, and Strangeness of Insect Societies* (New York: W. W. Norton, 2008).

16. Herbert Spencer, *The Principles of Biology* (New York: D. Appleton, 1866), 250.

17. Daniel Todes, *Darwin without Malthus: The Struggle for Existence in Russian Evolutionary Thought* (New York: Oxford University Press, 1989).

18. J. S. Schwartz, "Robert Chambers and Thomas Henry Huxley, Science Correspondents: The Popularization and Dissemination of Nineteenth Century Natural Science," *Journal of the History of Biology* 32 (1999): 366.

19. Todes, *Darwin without Malthus,* 126.

20. Charles Darwin, *The Autobiography of Charles Darwin, 1809–1882* (New York: W. W. Norton, 1958), 80.

21. Petr A. Kropotkin, *Mutual Aid: A Factor in Evolution* (New York: McClure Philips, 1902), 9.

22. Thomas H. Huxley, "The Struggle for Existence in Human Society," in *Evolution and Ethics,* ed. Thomas H. Huxley (London: Macmillan, 1894), 199–200.

23. Ibid., 211–212.

24. Petr Kropotkin, "The Theory of Evolution and Mutual Aid," *Nineteenth Century and After* 67 (January 1910): 87.

25. Petr Kropotkin, "The Direct Action of Environment on Plants," *Nineteenth Century and After* 68 (1910): 61, 75.

26. Kropotkin, "Theory of Evolution and Mutual Aid," 105–106.

27. Petr Kropotkin, "The Theory of Evolution and Mutual Aid," *Nineteenth Century and After* 67 (January 1910): 105.

28. Kropotkin, "Direct Action of the Environment and Evolution," 70–71.

29. Todes, *Darwin without Malthus*, 141.

30. Mark B. Adams, "The Founding of Population Genetics: Contributions of the Chetverikov School, 1924–1934," *Journal of the History of Biology* 1 (1968): 37–38.

31. Ibid., 24.

32. This characterization of Kropotkin's confusion was initially pointed out to me by Frederick Churchill.

33. Jack Morrell, *Science at Oxford, 1914–1939: Transforming an Arts University* (Oxford: Clarendon Press, 1997), 5.

34. Ibid., 269–286.

35. Ibid., 287.

36. Ibid., 283.

37. G. D. H. Carpenter and E. B. Ford, *Mimicry* (London: Methuen, 1933).

Chapter Three

1. V. C. Wynne-Edwards, "Backstage and Upstage with 'Animal Dispersion,'" in *Leaders in the Study of Animal Behavior: Autobiographical Perspectives,* ed. Donald Dewsbury (London: Associated University Presses, 1985), 488.

2. Ibid.

3. Wynne-Edwards Collection, Queen's University Archive, box 12, file 10, diary 1921, 25.

4. For an excellent discussion of the role of natural theology, especially in Britain, see J. H. Brooke, *Science and Religion: Some Historical Perspectives* (Cambridge: Cambridge University Press, 1991), chaps. 6–8.

5. Wynne-Edwards, "Backstage and Upstage with 'Animal Dispersion,'" 488–489.

6. Wynne-Edwards, "Backstage and Upstage with "'Animal Dispersion,'" 489.

7. Ibid., 490.

8. Peter Crowcroft, *Elton's Ecologists: A History of the Bureau of Animal Population* (Chicago: University of Chicago Press, 1991), 2.

9. Ibid., 3–4.

10. Wynne-Edwards Collection, Queen's University Archive, box 2, file 6.

11. For biographical information on Carr-Saunders, see *Dictionary of National Biography,* s.v. "Carr-Saunders, Alexander."

12. Jack Morrell, "The Non-medical Sciences, 1914–1939," in *The History of the University of Oxford,* vol. 8, ed. B. Harrison (Oxford: Clarendon Press, 1994), 139–163.

13. Wynne-Edwards Collection, Queen's University Archive, box 2, file 19.

14. Ibid.

15. Wynne-Edwards Collection, Queen's University Archive, box 12, file 11, diary 1924, 131–132.

16. V. C. Wynne-Edwards, "The Behaviour of Starlings in Winter: An Investigation of the Diurnal Movements and Social Roosting-Habit," *British Birds* 23 (1929): 140.

17. Ibid., 177–178.

18. Wynne-Edwards, "Backstage and Upstage with 'Animal Dispersion,'" 494.

19. V. C. Wynne-Edwards, "On the Habits and Distribution of Birds on the North Atlantic," *Proceedings of the Boston Society of Natural History* 40, 4 (1935): 239.

20. Charles Darwin, *On the Origin of Species* (Cambridge, MA: Harvard University Press, 1859), 62–63.

21. Petr A. Kropotkin, *Mutual Aid: A Factor in Evolution* (New York: McClure Philips, 1902).

22. V. C. Wynne-Edwards, "Zoology of the Baird Expedition," *Auk* 69 (1952): 384.

23. Wynne-Edwards Collection, Queen's University Archive, box 3, file 11, diary 1922, 68.

24. Wynne-Edwards Collection. Queen's University Archive, box 3, file 11, diary 1937, 70–77.

25. Wynne-Edwards, "Backstage and Upstage with 'Animal Dispersion,'" 499–500.

26. Patrick D. Baird, "Baffin Expedition 1950," *Canadian Geographical Journal* 40 (1950): 212–223.

27. V. C. Wynne-Edwards, "The Nature of Subspecies," *Scottish Naturalist* 60 (1948): 195–196.

Chapter Four

1. David L. Lack, "My Life as an Amateur Ornithologist," *Ibis* 115 (1973): 421–431.

2. For an excellent description of the development of Gause's principle of competitive exclusion, see Sharon E. Kingsland, *Modeling Nature: Episodes in the History of Population Ecology* (Chicago: University of Chicago Press, 1985), 146–162.

3. For a recent reassessment of Galápagos finch evolution that is closer to Darwin's interpretation of ecological divergence, see Jonathon Weiner, *The Beak of the Finch: A Story of Evolution in Our Time* (New York: Knopf, 1994). See also Peter Grant and Rosemary Grant, *How and Why Species Multiply: The Radiation of Darwin's Finches* (Princeton, NJ: Princeton University Press, 2007).

4. Kingsland, *Modeling Nature,* 169.

5. Peter Crowcroft, *Elton's Ecologists: A History of the Bureau of Animal Population* (Chicago: University of Chicago Press, 1991), 138.

6. See "In Appreciation: Tributes to David Lack," *Ibis* 115 (1973): 431–441.

7. For discussion and analysis of the modern synthesis, see M. R. Dietrich, "Paradox and Persuasion: Negotiating the Place of Molecular Evolution within Evolutionary Biology," *Journal of the History of Biology* 31 (1998): 85–111; Stephen J. Gould, The Hardening of the Modern Synthesis," in *Dimensions of Darwinism,* ed. Marjorie Grene (Cambridge: Cambridge University Press, 1983), 71–93; Marjorie Grene, ed., *Dimensions of Darwinism: Themes and Counterthemes in Twentieth Century Evolutionary Theory* (Cambridge: Cambridge University Press, 1983); Ernst Mayr and W. B. Provine, eds., *The Evolutionary Synthesis: Perspectives on the Unification of Biology* (Cambridge, MA: Harvard University Press, 1980); Ernst Mayr, "What Was the Evolutionary Synthesis?" *Trends in Ecology and Evolution* 8 (1993): 31–34; Vasiliki Betty Smocovitis, *Unifying Biology: The Evolutionary Synthesis and Evolutionary Biology* (Princeton, NJ: Princeton University Press, 1996).

8. Theodosius Dobzhansky, *Genetics and the Origin of Species,* ed. Leslie C. Dunn (New York: Columbia University Press, 1937).

9. Ibid., 127.

10. Ibid., 120.

11. Sewall Wright, "Fisher's Theory of Dominance," *American Naturalist* 63 (1929): 274–279; Sewall Wright, "Evolution in Mendelian Populations," *Genetics* 16 (1931): 97–159; Sewall Wright, *The Roles of Mutation, Inbreeding, Crossbreeding and Selection in Evolution* (Ithaca, NY: Cornell University Press, 1932), 356–366.

12. Julian S. Huxley, *The Individual in the Animal Kingdom* (Cambridge: Cambridge University Press, 1912); Julian Huxley, *Evolution: The Modern Synthesis* (London: Allen and Unwin, 1942); Ernst Mayr, *Systematics and the Origin of Species* (New York: Columbia University Press, 1942); Ernst Mayr, *Animal Species and Evolution* (Cambridge, MA: Harvard University Press, 1963); George Gaylord Simpson, *Tempo and Mode in Evolution* (New York: Columbia University Press, 1944).

13. Mayr, *Systematics and the Origin of Species*, 190.

14. Gould, "Hardening of the Modern Synthesis," 71–93.

15. Wynne-Edwards Collection, Queen's University Archive, box 2, file 11.

16. Ibid.

17. Wynne-Edwards Collection, Queen's University Archive, box 2, file 15.

18. V. C. Wynne-Edwards, "The Dynamics of Animal Populations," *Discovery: A Monthly Popular Journal of Knowledge* 16 (1955): 434.

19. Ibid., 433.

20. Ibid., 434.

21. Ibid., 436.

22. Wynne-Edwards Collection, Queen's University Archive, box 2, file 2.

23. V. C. Wynne-Edwards, "Low Reproductive Rates in Birds, Especially Sea-Birds," *Acta of the Eleventh International Congress of Ornithology,* 545–546; emphasis added.

24. Ibid., 547.

25. Wynne-Edwards Collection, Queen's University Archive, box 2, file 6.

26. Wynne-Edwards Collection, Queen's University Archive, box 2, file 4.

27. Wynne-Edwards Collection, Queen's University Archive, box 1, file 16.

28. Crowcroft, *Elton's Ecologists,* 41.

29. Henry N. Southern, "Mortality and Population Control," *Ibis* 101 (1959): 429–436.

30. Ibid., 435.

31. V. C. Wynne-Edwards, "The Control of Population Density through Social Behaviour: A Hypothesis," *Ibis* 101 (1959): 440.

32. J. A. Simpson and E. S. C. Weiner, eds., *The Oxford English Dictionary* (Oxford: Clarendon Press, 1991), 328.

33. Wynne-Edwards, "Control of Population Density through Social Behaviour," 437.

34. Ibid., 440.

35. Wynne-Edwards Collection, Queen's University Archive, box 2, file 15, 1.

36. Ibid.
37. Ibid.
38. Ibid.
39. Ibid., 2.
40. Wynne-Edwards Collection, Queen's University Archive, box 1, file 16.
41. Wynne-Edwards Collection, Queen's University Archive, box 2, file 15.
42. Wynne-Edwards Collection, Queen's University Archive, box 2, file 8.
43. Ibid.

Chapter Five

1. See, for example, Bernard J. Boelen, *Symposium on Evolution, Held in Commemoration of the Centenary of Charles Darwin's "The Origin of Species"* (Pittsburgh, PA: Duquesne University Press, 1959); G. W. Leeper, *The Evolution of Living Organisms: A Symposium to Mark the Centenary of Darwin's "Origin of Species" and of the Royal Society of Victoria,* Melbourne, December 1959 (Parkville: Melbourne University Press, 1962); Sol Tax, ed., *Evolution after Darwin,* University of Chicago Centennial (Chicago: University of Chicago Press, 1960); P. J. Wanstall, ed., *A Darwin Centenary: The Report of the Conference Held by the Botanical Society of the British Isles in 1959 to Mark the Centenary of the Publication of "The Origin of Species"* (London: Botanical Society of the British Isles, 1961). For the most recent collection of scholarship on the synthesis, see Joe Cain and Michael Ruse, eds., *Descended from Darwin: Insights into the History of Evolutionary Studies, 1900–1970* (Philadelphia, PA: American Philosophical Society Press, 2009).

2. Theodosius D. Dobzhansky, "Nothing in Biology Makes Sense Except in the Light of Evolution," *American Biology Teacher* 35 (1973): 125–129.

3. Joel Hagen, *An Entangled Bank: The Origins of Ecosystem Ecology* (New Brunswick, NJ: Rutgers University Press, 1992).

4. See M. R. Dietrich, "Paradox and Persuasion: Negotiating the Place of Molecular Evolution within Evolutionary Biology," *Journal of the History of Biology* 31 (1998): 85–111.

5. See Ernst Mayr, "The New versus the Classical in Science," *Science* 141 (1963): 763, and Theodosius D. Dobzhansky, "Are Naturalists Old-Fashioned?" *American Naturalist* 100 (1966): 541–550. Both these articles express the authors' concern about the overemphasis on molecular biology to the detriment of more traditional studies. This concern for the reputation of "classical" biology may have contributed to the skepticism with which Wynne-Edwards's book was met. Although both Lack and Wynne-Edwards would have agreed with Dobzhansky and Mayr on this point, I argue that Wynne-Edwards's work would have caused concern for the others, because although it was indeed "classical" and organismal, it did not hew to the developing neo-Darwinian orthodoxy.

6. V. C. Wynne-Edwards, *Animal Dispersion in relation to Social Behavior* (Edinburgh: Oliver and Boyd, 1962), 19.

7. Ibid., v.

8. Wynne-Edwards Collection, Queen's University Archive, box 2, file 11.

9. Wynne-Edwards Collection, Queen's University Archive, box 1, file 13.

10. Wynne-Edwards, *Animal Dispersion,* v.

11. Ibid.

12. Charles Darwin, *On the Origin of Species* (Cambridge, MA, Harvard University Press, 1859), 488.

13. James D. Watson and Francis Crick, "A Structure for Deoxyribose Nucleic Acid," *Nature* 171 (1953): 737.

14. Wynne-Edwards, *Animal Dispersion,* 14.

15. Ibid., 20.

16. Ibid., 493.

17. Ibid., 22.

18. Erik Angner, "The History of Hayek's Theory of Cultural Evolution," *Studies in History and Philosophy of Biological and Biomedical Sciences* 33 (2002): 695–718.

19. See, for example, the work of another Austrian economist, Ernst Fehr, including Herbert Gintis, Samuel Bowles, Ernst Fehr, et al., eds., *Moral Sentiments and Material Interests: The Foundations of Cooperation in Economic Life*, Economic Learning and Social Evolution (Cambridge, MA: MIT Press, 2005), and H. Bernhard, U. Fischbacher, and E. Fehr, "Parochial Altruism in Humans," *Nature* 442 (2006): 912–915.

20. Wynne-Edwards, *Animal Dispersion,* 2.

21. For an excellent analysis of the work of Allee and the Chicago ecology program, see Gregg Mitman, *The State of Nature: Ecology, Community and American Social Thought, 1900–1950* (Chicago: University of Chicago Press, 1992).

22. Wynne-Edwards, *Animal Dispersion,* 129.

23. Ibid., 132.

24. Ibid.

25. See V. C. Wynne-Edwards, "On the Waking-Time of the Nightjar," *Journal of Experimental Biology* 7 (1929): 241–247, and V. C. Wynne-Edwards, "The Behaviour of Starlings in Winter: An Investigation of the Diurnal Movements and Social Roosting-Habit," *British Birds* 23 (1929): 138–153, 170–180.

26. Wynne-Edwards Collection, Queen's University Archive, box 4, file 13.

27. Charles Elton, "Self-Regulation of Animal Populations," *Nature* 197 (1963): 634.

28. Ibid.

29. Wynne-Edwards Collection, Queen's University Archive, box 2, file 6.

30. For an excellent collection of essays on iconoclastic biologists, see Oren

Harman and Michael R. Dietrich, eds., *Rebels, Mavericks, and Heretics in Biology* (New Haven, CT: Yale University Press, 2008).

31. Elton, "Self-Regulation of Animal Populations," 634.

32. For an excellent treatment of the development of population biology and the most significant hypotheses, see Dennis Chitty, *Do Lemmings Commit Suicide? Beautiful Hypotheses and Ugly Facts* (New York: Oxford University Press, 1996).

33. Elton, "Self-Regulation of Animal Populations," 634.

34. F. W. Braestrup, "Special Review," *Oikos* 14 (1963): 113–120.

35. Ibid., 113.

36. C. M. Waters and B. L. Bassler, "Quorum Sensing: Cell to Cell Communication in Bacteria," *Annual Review of Cell and Developmental Biology* 21 (2005): 319–346.

37. Braestrup, "Special Review," 117.

38. Ibid., 118.

39. Ibid., 119.

40. J. A. King, "Social Behavior and Population Homeostasis," *Ecology* 46 (1965): 210.

Chapter Six

1. David Cort, "The Glossy Rats: A Review of *Animal Dispersion in relation to Social Behavior*," *Nation* 197 (November 16, 1963): 327.

2. Ibid.

3. Ibid.

4. "The Nature of Social Life," *Times Literary Supplement* (December 14, 1962): 967; all quotations are from this page.

5. Ibid.

6. Ibid.

7. Ibid.

8. Ibid.

9. Ibid.

10. Ibid.

11. Ibid.

12. Ibid.

13. Letter from Charles Sibley to David Lack, David Lack Papers, Alexander Library, Edward Grey Institute, box 10, file 223.

14. Letter from Charles Sibley to David Lack, David Lack Papers, Alexander Library, Edward Grey Institute, box 10, file 223.

15. Robert MacArthur to David Lack, David Lack Papers, Alexander Library, Edward Grey Institute, box 10, file 223.

16. *TLS* review (1962) as cited in David Lack, *Population Studies of Birds* (Oxford: Clarendon Press, 1966), 311.

17. Lack, *Population Studies of Birds,* 311. With regard to the status of density dependence theories in population studies since Nicholson, see Dennis Chitty, *Do Lemmings Commit Suicide? Beautiful Hypotheses and Ugly Facts* (New York: Oxford University Press, 1996).

18. Wynne-Edwards Collection, Queen's University Archive, box 15, file 12.

19. Lack, *Population Studies of Birds,* 8; emphasis added.

20. Ibid., 121–122. See also Christopher M. Perrins, "Survival in the Great Tit, *Parus major*," *Proceedings of the International Ornithological Congress* 13 (1963): 717–728; Christopher M. Perrins, "Population Fluctuations and Clutch-Size in the Great Tit *Parus major*," *Journal of Animal Ecology* 34 (1965): 601–647; and D. W. Snow, "The Breeding of the Blackbird, *Turdus merula*, at Oxford," *Ibis* 100 (1958): 1–30.

21. V. C. Wynne-Edwards, *Animal Dispersion in relation to Social Behavior* (Edinburgh: Oliver and Boyd, 1962), 568; emphasis added.

22. Lack, *Population Studies of Birds,* 222.

23. Ibid., 228.

24. Ibid., 257.

25. Ibid., 261.

26. Ibid., 262; emphasis added.

27. Ibid., 275.

28. Thomas Park as cited in Lack, *Population Studies of Birds,* 280.

29. Gregg Mitman, *The State of Nature: Ecology, Community and American Social Thought, 1900–1950* (Chicago: University of Chicago Press, 1992). See also Gregg Mitman, "From Population to Society: The Cooperative Metaphors of W. C. Allee and A. E. Emerson," *Journal of the History of Biology* 21 (1988): 173–194.

30. George C. Williams, *Adaptation and Natural Selection: A Critique of Some Current Evolutionary Thought,* rev. ed. (1966; Princeton, NJ: Princeton University Press, 1996), ix.

31. Ibid., x.

32. Ibid., 4–5.

33. Ibid., 108.

34. Ibid., 160.

35. Ibid., xii.

36. Elliott Sober and David Sloan Wilson, *Unto Others: The Evolution and Psychology of Unselfish Behavior* (Cambridge, MA: Harvard University Press, 1998); see also George C. Williams, *Natural Selection: Domains, Levels, Challenges* (New York: Oxford University Press, 1992), and W. D. Hamilton, *Narrow Roads of Gene Land: The Collected Papers of W. D. Hamilton,* vol. 1, *Evolution of Social Behavior* (Oxford: W. H. Freeman, 1996).

37. E. A. Lloyd, "Altruism Revisited," *Quarterly Review of Biology* 74 (1999): 447.

38. Letter from George C. Williams to David Lack, David Lack Papers, Alexander Library, Edward Grey Institute, box 11, file 268.

39. Richard W. Burkhardt, "On the Emergence of Ethology as a Scientific Discipline," *Conspectus of History* 1 (1981): 63.

40. Tom Gieryn, "Boundaries of Science," in *Handbook of Science, Technology and Society*, ed. S. Jasanoff et al. (Beverly Hills, CA: Sage, 1994), 393–443.

41. Burkhardt, "On the Emergence of Ethology as a Scientific Discipline," 65.

42. For the most compelling and comprehensive treatment of the development of the field of ethology, and especially the relationship of its two most significant founders, see Richard W. Burkhardt, *Patterns of Behavior: Konrad Lorenz, Niko Tinbergen and the Founding of Ethology* (Chicago: University of Chicago Press, 2005).

43. John R. Krebs, "The Book That Most . . ." *Biologist* 39 (1992): 217.

44. F. J. Ebling and D. M. Stoddart, eds., *Population Control by Social Behaviour,* Symposia of the Institute of Biology (London: Institute of Biology, 1978), vii.

45. Ibid., x.

46. Richard W. Burkhardt, "Huxley and the Rise of Ethology," in *Julian Huxley: Biologist and Statesman of Science,* ed. C. Kenneth Waters and Albert Van Helden (Houston, TX: Rice University Press, 1992), 134–135.

47. Mitman, *State of Nature,* 72.

48. Burkhardt, "Huxley and the Rise of Ethology," 135.

49. Mitman, *State of Nature,* 111.

50. Burkhardt, "Huxley and the Rise of Ethology," 148.

51. V. C. Wynne-Edwards, "The Control of Population Density through Social Behaviour: A Hypothesis," *Ibis* 101 (1959): 440.

52. Ibid., 437.

53. Konrad Lorenz, "Analogy as a Source of Knowledge," *Science* 185 (1974): 229–234.

54. Konrad Lorenz, "Part and Parcel in Animal and Human Societies: A Methodological Discussion," in *Studies in Animal and Human Behavior,* vol. 2, ed. and trans. R. Martin (Cambridge, MA: Harvard University Press, 1971), 115–195.

55. Lorenz, "Part and Parcel in Animal and Human Societies," 129.

56. Wynne-Edwards, *Animal Dispersion,* 147.

57. Burkhardt, "Huxley and the Rise of Ethology," 145.

58. Wynne-Edwards, *Animal Dispersion,* v.

59. Charles Darwin, *On the Origin of Species* (Cambridge, MA: Harvard University Press, 1859), 488.

60. R. W. Burkhardt, "The Development of an Evolutionary Ethology," in *Evolution from Molecules to Men,* ed. D. Bendall (Cambridge: Cambridge University Press, 1983), 436–437.

61. Ibid., 437.

62. Konrad Lorenz, *Studies in Animal and Human Behavior* (Cambridge, MA: Harvard University Press, 1970), 218.

63. Julian S. Huxley, "Lorenzian Ethology," *Zeitschrift für Tierpsychologie* 20 (1963): 407.

64. Niko Tinbergen, "Behavior and Natural Selection," in *Ideas in Modern Biology,* ed. J. Moore (Garden City, NJ: Natural History Press, 1965), 521.

65. Ibid., 522.

66. Ibid., 530.

67. Ibid., 536.

68. Robert A. Hinde, "Ethology in relation to Other Disciplines," in *Leaders in the Study of Animal Behavior,* ed. Donald Dewsbury (London: Associated University Press, 1985), 194.

69. V. C. Wynne-Edwards, "Space Use and the Social Community in Animals and Men," in *Behavior and Environment: The Use of Space in Animals and Men,* ed. Aristide H. Esser (New York: Plenum Press, 1971), 270.

70. Konrad Lorenz, *On Life and Living: Konrad Lorenz in Conversation with Kurt Mündl* (New York: St. Martin's Press, 1990), 149.

71. Konrad Lorenz, *The Waning of Humaneness* (Boston: Little, Brown, 1987), 4.

72. V. C. Wynne-Edwards, "A Rationale for Group Selection," *Journal of Theoretical Biology* 162 (1993): 20–21.

73. Peter H. Klopfer, *Politics and People in Ethology* (London: Associated University Press, 1999), 95.

74. Ibid., 99.

75. There is correspondence between David Sloan Wilson and Wynne-Edwards in the collection at Queen's University from the late 1970s until 1982. The influence on Michael Wade was described in personal communications with him between 2000 and 2008.

76. Mitman, *State of Nature,* 208.

77. Edward Goldsmith et al., *Blueprint for Survival* (Boston: Houghton Mifflin, 1972).

78. From the text of the original warrant accessed on the Royal Commission on Environmental Pollution Web site <http://www.rcep.org.uk/about/warrant2.htm>. Accessed May 22, 2008.

79. Wynne-Edwards, "Space Use and the Social Community," 276.

80. Ibid., 278

81. Ibid., 279.

82. Sewall Wright, *Evolution and the Genetics of Populations,* vol. 2, *The Theory of Gene Frequencies* (Chicago: University of Chicago Press, 1969), 129.

Chapter Seven

1. Elliott Sober and Richard Lewontin, "Artifact, Cause, and Genic Selection," *Philosophy of Science* 49 (1982): 157–180.

2. Ernst Mayr, *Animal Species and Evolution* (Cambridge, MA: Harvard University Press, 1963), 184.

3. William D. Hamilton, "The Genetical Evolution of Social Behavior, I and II," *Journal of Theoretical Biology* 7 (1964): 1–52.

4. Robert Trivers, "The Evolution of Reciprocal Altruism," *Quarterly Review of Biology* 46 (1971): 35–57.

5. John Maynard Smith, "Group Selection and Kin Selection," *Nature* 201 (1964): 1146.

6. Ibid. As Michael Wade pointed out in his 1978 article, Maynard Smith had given all power to individual selection, in that any group with even one selfish allele became nonaltruistic. Only groups with no selfish alleles, those created by drift, persisted.

7. V. C. Wynne-Edwards, "Group Selection and Kin Selection: Response to Maynard-Smith," *Nature* 201 (1964): 1147.

8. Ibid.

9. Ibid.

10. Gregory B. Pollock, "Suspending Disbelief—of Wynne-Edwards and His Reception," *Journal of Evolutionary Biology* 2 (1989): 207.

11. Personal communication with Michael Wade (2000) and Gregory Pollock (2007).

12. Michael J. Wade, "An Experimental Study of Kin Selection," *Evolution* 34 (1980): 844–855; Michael J. Wade, "Kin Selection: Its Components," *Science* 210 (1980): 665–667.

13. Elliott Sober and Donald Sloan Wilson, *Unto Others: The Evolution and Psychology of Unselfish Behavior* (Cambridge, MA: Harvard University Press 1998), 71. This point was also made by Michael Wade, "A Critical Review of the Models of Group Selection," *Quarterly Review of Biology* 53 (1978): 101–114.

14. Wynne-Edwards Collection, Queen's University Archive, box 15, file 12, Notebook Series, Third Black Book, 71–72.

15. Theodosius D. Dobzhansky, "Mendelian Populations and Their Evolution," in *Genetics in the Twentieth Century,* ed. L. C. Dunn (New York: Macmillan, 1951), 573–589.

16. Wynne-Edwards Collection, Queen's University Archive, box 15, file 12, Notebook Series, Third Black Book, 73.

17. Social Science Research Council Records, Rockefeller Archive Center, Sleepy Hollow, New York, accession 2, box 189, folder 2151.

18. V. C. Wynne-Edwards, "Backstage and Upstage with *Animal Dispersion*," in *Leaders in the Study of Animal Behavior: Autobiographical Perspectives,* ed. Donald Dewsbury (London: Associated University Presses, 1985), 510.

19. See Arthur L. Caplan, ed., *The Sociobiology Debate: Readings on Ethical and Scientific Issues* (New York: Harper and Row, 1978); Stephen Jay Gould and Richard Lewontin, "The Spandrels of San Marco and the Panglossian Paradigm,"

Proceedings of the Royal Society of London 205 (1978): 581–598; Howard L. Kaye, *The Social Meaning of Modern Biology: From Social Darwinism to Sociobiology* (New Haven, CT: Yale University Press, 1986); Philip Kitcher, *Vaulting Ambition: Sociobiology and the Quest for Human Nature* (Cambridge, MA: MIT Press, 1985); Richard Lewontin, Steven Rose, and Leon J. Kamin, *Not in Our Genes: Biology, Ideology and Human Nature* (New York: Pantheon Books, 1984); Ullica Segerstråle, *Defenders of the Truth: The Battle for Science in the Sociobiology Debate and Beyond* (Oxford: Oxford University Press, 2000).

20. Edward O. Wilson, *Naturalist* (Washington, DC: Island Press, 1994), 330.

21. Ibid., 333.

22. Edward O. Wilson, *Sociobiology: The New Synthesis* (Cambridge, MA, Belknap Press of Harvard University Press, 1975), 110; emphasis added.

23. See G. E. Allen, "Naturalists and Experimentalists: The Genotype and the Phenotype," *Studies in History of Biology* 3 (1979): 179–209, and Richard W. Burkhardt, "Ethology, Natural History, the Life Sciences, and the Problem of Place," *Journal of the History of Biology* 32 (1999): 489–509.

24. V. C. Wynne-Edwards, "*The Genetics of Altruism*—Review," *Social Science and Medicine* 16 (1982): 1096.

25. Segerstråle, *Defenders of the Truth,* 85.

26. Edward O. Wilson and Bert Hölldobler, "Eusociality: Origin and Consequences," *Proceedings of the National Academy of Sciences* 102 (2005): 13367–13371.

27. David Sloan Wilson, and Edward O. Wilson, "Rethinking the Theoretical Foundation of Sociobiology," *Quarterly Review of Biology* 82 (2007): 327–348.

28. Wynne-Edwards Collection, Queen's University Archive, box 15, file 12, Notebook Series, Fourth Black Book, 37.

29. Richard Dawkins, *The Selfish Gene,* new ed. (1976; London: Oxford University Press, 1989), ix–x.

30. Ibid., ix.

31. V. C. Wynne-Edwards, "Intrinsic Population Control: An Introduction," in *Population Control by Social Behavior,* ed. F. J. Ebling and D. M. Stoddart (London: Institute of Biology, 1978), 19.

32. Dawkins, *Selfish Gene,* 297. Here Dawkins is referring to *Evolution through Group Selection,* which had been published just a few years before the revised edition of *The Selfish Gene.*

33. Wynne-Edwards Collection, Queen's University Archive, box 15, file 12, Notebook Series, Fourth Black Book, 79.

34. Richard Lewontin, "The Units of Selection," *Annual Review of Ecology and Systematics* 1 (1970): 1–18.

35. See especially David L. Hull, "Are Species Really Individuals?" *Systematic Zoology* 25 (1976): 174–191, and David Hull, "Individuality and Selection," *Annual Review of Ecology and Systematics* 11 (1980): 311–332.

36. See William Wimsatt, "Reductionist Research Strategies and Their Biases in the Units of Selection Controversy," in *Scientific Discovery: Case Studies,* ed. Thomas Nickles (Dordrecht: Reidel, 1980), 213–259, and William Wimsatt, "Units of Selection and the Structure of the Multi-level Genome," *Proceedings of the Philosophy of Science Association* 2 (1981): 122–183.

37. For an excellent collection of many of the most important papers in the units of selection debate, see Robert N. Brandon and Richard M. Burian, eds., *Genes, Organisms, Populations: Controversies over the Units of Selection* (Cambridge, MA: MIT Press, 1984). Other important contributions include Robert N. Brandon, *Adaptation and Environment* (Princeton, NJ: Princeton University Press, 1990); Robert N. Brandon, *Concepts and Methods in Evolutionary Biology* (Cambridge: Cambridge University Press, 1996); Kim Sterelny and Philip Kitcher, "The Return of the Gene," *Journal of Philosophy* 85 (1988): 339–361; Kim Sterelny, "The Return of the Group," *Philosophy of Science* 63 (1996): 562–584; and Michael Ruse, "Charles Darwin and Group Selection," *Annals of Science* 37 (1980): 615–630.

38. See, for example, Elliott Sober, "Holism, Individualism and the Units of Selection," *Proceedings of the Philosophy Association 1980,* 2 (1981): 93–121, and Elliott Sober, *The Nature of Selection: Evolutionary Theory in Philosophical Focus* (Chicago: University of Chicago Press, 1984). His coauthored articles include Sober and Lewontin, "Artifact, Cause and Genic Selection," 157–180. Collaborations with David Sloan Wilson include David Sloan Wilson and Elliott Sober, "Reviving the Superorganism," *Journal of Theoretical Biology* 136 (1989): 337–356; David Sloan Wilson and Elliott Sober, "Reintroducing Group Selection to the Human Sciences," *Behavioral and Brain Sciences* 17 (1994): 585–654; and Sober and Wilson, *Unto Others.*

39. Elisabeth A. Lloyd, *The Structure and Confirmation of Evolutionary Theory* (Princeton, NJ: Princeton University Press, 1988); Elisabeth Lloyd and Stephen Jay Gould, "Species Selection on Variability," *Proceedings of the National Academy of Sciences* 90 (1993): 595–599.

40. James R. Griesemer and Michael J. Wade, "Laboratory Models, Causal Explanation, and Group Selection," *Biology and Philosophy* 3 (1988): 67–96.

41. Elisabeth A. Lloyd, "Units and Levels of Selection: An Anatomy of the Units of Selection Debates," in *Thinking about Evolution: Historical, Philosophical, and Political Perspectives,* ed. Rama S. Singh et al. (Cambridge: Cambridge University Press, 2001), 267–291.

42. Ibid., 272.

43. V. C. Wynne-Edwards, *Evolution through Group Selection* (Oxford: Blackwell Scientific Publications, 1986), 200.

44. Ibid., 203.

45. Ibid., 204.

46. Ibid., 206–207.

47. Ibid., 206.

48. Ibid., 207.

49. Letter from Sewall Wright to V. C. Wynne-Edwards, March 11, 1983, Wynne-Edwards Collection, Queen's University Archive, box 16, file 2.

50. Letter from V. C. Wynne-Edwards to Michael Wade. Wade gave me a copy of this letter from his files.

51. Michael J. Wade, "*Evolution through Group Selection*," *Evolution* 42 (1988): 1116.

52. Mark Ridley, "*Evolution through Group Selection*—Review," *Ethology* 74 (1987): 260–261.

53. Ibid., 261.

54. Mary Jane West-Eberhard, "*Evolution through Group Selection*—Review," *Journal of Genetics* 65 (1986): 213–217; emphasis added.

55. Wynne-Edwards Collection, Queen's University Archive, box 4, file 14; letter from Roy Taylor, December 8, 1986.

56. Wynne-Edwards Collection, Queen's University Archive, box 4, file 14; letter from Wynne-Edwards, December 17, 1986.

57. Ibid., 2.

58. Wynne-Edwards, *Evolution through Group Selection,* 84.

59. Wynne-Edwards Collection, Queen's University Archive, box 4, file 14; letter from Wynne-Edwards, December 17, 1986, 2.

60. Wynne-Edwards Collection, Queen's University Archive, box 4, file 14; letter to Roy Taylor, January 5, 1987.

61. Wynne-Edwards Collection, Queen's University Archive, box 4, file 14; letter to Adam Watson, January 19, 1987.

62. Wynne-Edwards Collection, Queen's University Archive, box 4, file 14; letter to Robert Campbell, June 29, 1984; emphasis added.

Chapter Eight

1. Donald A. Dewsbury, ed., *Leaders in the Study of Animal Behavior: Autobiographical Perspectives* (Lewisburg, PA: Bucknell University Press, 1985).

2. Wynne-Edwards Collection, Queen's University Archive, box 5, file 6; reply to Perot Walker, October 22, 1987.

3. Wynne-Edwards Collection, Queen's University Archive, box 2, file 20; manuscript of response from Heinen and Low to 1991 *Ecologist* article, 1.

4. Ibid.

5. Ibid., 3.

6. V. C. Wynne-Edwards, "Ecology Denies Neo-Darwinism," *Ecologist* 21 (1991): 138; emphasis added.

7. Wynne-Edwards Collection, Queen's University Archive, box 2, file 20; manu-

script of Heinen and Low's response to Wynne-Edwards's 1991 *Ecologist* article, 7.

8. Wynne-Edwards Collection, Queen's University Archive, box 2, file 20.

9. J. Papworth, "Mass Society and the Pheasant," *Ecologist* 21 (1991): 232.

10. M. Begg, "Self-Interest for the Good of All," *Ecologist* 21 (1991): 272.

11. Wynne-Edwards Collection, Queen's University Archive, box 1, file 1; letter to Lil Lecock.

12. Wynne-Edwards Collection, Queen's University Archive, box 1, file 1; letter to *Nature* editor John Maddox.

13. Ibid.

14. Wynne-Edwards Collection, Queen's University Archive, box 2, file 19; letter to *Nature* editor, June 14, 1989.

15. Wynne-Edwards Collection, Queen's University Archive, box 2, file 19; letter from John Maddox, June 30, 1989.

16. Wynne-Edwards Collection, Queen's University Archive, box 2, file 19; letter from John Maddox, October 23, 1989.

17. Wynne-Edwards Collection, Queen's University Archive, box 2, file 20; letter from Felicity Huntingford, February 6, 1990; Referee's Report B.

18. Wynne-Edwards Collection, Queen's University Archive, box 2, file 20; letter from J. Z. Young, July 19, 1991.

19. Wynne-Edwards Collection, Queen's University Archive, box 2, file 20l; reply to J. Z. Young, July 31, 1991.

20. Personal communication (1999).

21. David Jenkins and Adam Watson, "Obituary: Vero Copner Wynne-Edwards (1906–1997)," *Ibis* 139 (1997): 415–418; I. Newton, "Vero Copner Wynne-Edwards C.B.E." *Biographical Memoirs of the Fellows of the Royal Society* 44 (1998): 473–484.

22. Wynne-Edwards Collection, Queen's University Archive, box 4, file 1; letter to Robert May, October 9, 1993.

23. Jenkins and Watson, "Obituary," 418.

24. Frederick B. Churchill, "The History of Embryology as Intellectual History," *Journal of the History of Biology* 3 (1970): 179.

25. Robert P. McIntosh, "Citation Classics of Ecology," *Quarterly Review of Biology* 64 (1989): 31–49.

26. Ibid., 37.

27. There is correspondence between David Sloan Wilson and Wynne-Edwards in the collection at Queen's University from the late 1970s until 1982. The influence on Michael Wade was described in personal communications with him between 2000 and 2008.

28. Norwood Russell Hanson, "Observation," in *Introductory Readings in the Philosophy of Science,* ed. E. D. Klemke, Robert Hollinger, and A. David Kline (Buffalo, NY: Prometheus Books, 1988).

29. Thomas H. Huxley, "The Struggle for Existence in Human Society," in *Evolution and Ethics,* by Thomas H. Huxley, 9:195–236 (London: Macmillan, 1893).

30. Charles Darwin, *On the Origin of Species* (Cambridge, MA: Harvard University Press, 1859), 62–63.

31. Ayelet Shavit, "Shifting Values Partly Explain the Debate over Group Selection," *Studies in the History and Philosophy of Biological and Biomedical Sciences* 35 (2004): 700.

32. Ibid., 705.

33. Ibid., 717.

34. See, for example, S. A. West, A. S. Griffin, and A. Gardener, "Social Semantics: Altruism, Cooperation, Mutualism, Strong Reciprocity, and Group Selection," *Journal of Evolutionary Biology* 20 (2007): 415–432, and David Sloan Wilson, "Social Semantics: Toward a Genuine Pluralism in the Study of Social Behavior," *Journal of Evolutionary Biology* 21 (2008): 368–373.

35. David L. Hull, *Science as a Process: An Evolutionary Account of the Social and Conceptual Development of Science* (Chicago: University of Chicago Press, 1988), 481.

36. On the evolution of sex, see George C. Williams, *Natural Selection: Domains, Levels, and Challenges* (New York: Oxford University Press, 1992). On the evolution of social behavior, see Edward O. Wilson and Bert Hölldobler, "Eusociality: Origin and Consequences," *Proceedings of the National Academy of Sciences* 102 (2005): 13367–13371, and David Sloan Wilson and Edward O. Wilson, "Rethinking the Theoretical Foundations of Sociobiology," *Quarterly Review of Biology* 82 (2007): 327–348.

37. For primatology and anthropology see Christopher Boehm, *Hierarchy in the Forest: The Evolution of Egalitarian Behavior* (Cambridge, MA: Harvard University Press, 1999); Robert Boyd et al., "The Evolution of Altruistic Punishment," *Proceedings of the National Academy of Sciences* 100 (2003): 3531–3535; and Peter J. Richerson and Robert Boyd, *Not by Genes Alone: How Culture Transformed Human Evolution* (Chicago: University of Chicago Press, 2005). For economics see T. Bergstrom, "Evolution of Social Behavior: Individual and Group Selection," *Journal of Economic Perspectives* 16 (2002): 67–88, and S. Bowles and H. Gintis, "Social Capital and Community Governance," *Economic Journal* 112 (2002): F419–F436.

38. See Ben Kerr et al., "Local Migration Promotes Competitive Restraint in a Host-Pathogen 'Tragedy of the Commons,'" *Nature* 442 (2006): 75–78, and J. Werfel and Y. Bar-Yam, "The Evolution of Reproductive Restraint through Social Communication," *Proceedings of the National Academy of Sciences* 101 (2004): 11019–11024.

39. See Peter A. Corning, *Holistic Darwinism: Synergy, Cybernetics, and the Bioeconomics of Evolution* (Chicago: University of Chicago Press, 2005); Stephen Jay Gould, *The Structure of Evolutionary Theory* (Cambridge, MA: Harvard Uni-

versity Press, 2002); Samir Okasha, *Evolution and the Levels of Selection* (New York: Oxford University Press, 2006); and David Sloan Wilson, *Evolution for Everyone: How Darwin's Theory Can Change the Way We Think about Our Lives* (New York: Delacorte Press, 2007).

40. Gould, *Structure of Evolutionary Theory,* 548.

41. Okasha, *Evolution and the Levels of Selection,* 240.

Bibliography

Adams, M. B. (1968). "The Founding of Population Genetics: Contributions of the Chetverikov School, 1924–1934." *Journal of the History of Biology* 1 (1): 23–39.
———, ed. (1994). *The Evolution of Theodosius Dobzhansky: Essays on His Life and Thought in Russia and America.* Princeton, NJ: Princeton University Press.
Allee, W. C. (1943). "Where Angels Fear to Tread: A Contribution from General Sociology to Human Ethics." *Science* 97 (June 11): 517–525.
Allen, G. E. (1968). "Thomas Hunt Morgan and the Problem of Natural Selection." *Journal of the History of Biology* 1:113–140.
———. (1975). *Life Science in the Twentieth Century.* New York: John Wiley.
———. (1978). *Thomas Hunt Morgan: The Man and His Science.* Princeton, NJ: Princeton University Press.
———. (1979). "Naturalists and Experimentalists: The Genotype and the Phenotype." *Studies in History of Biology* 3:179–209.
———. (1981). "Morphology and Twentieth-Century Biology: A Response." *Journal of the History of Biology* 14 (1): 159–176.
Angner, E. (2002). "The History of Hayek's Theory of Cultural Evolution." *Studies in History and Philosophy of Biological and Biomedical Sciences* 33:695–718.
Baird, P. D. (1950). "Baffin Expedition 1950." *Canadian Geographical Journal* 40:212–223.
Ball, W. P. (1894). "Neuter Insects and Lamarckism." *Natural Science,* February 1894, 91–97.
Begg, M. (1991). "Self-Interest for the Good of All." *Ecologist* 21 (6): 272.
Benson, K. R., J. Maienschein, et al., eds. (1991). *The Expansion of American Biology.* New Brunswick, NJ: Rutgers University Press.
Bergstrom, T. (2002). "Evolution of Social Behavior: Individual and Group Selection." *Journal of Economic Perspectives* 16 (2): 67–88.
Bernhard, H., U. Fischbacher, and E. Fehr. (2006). "Parochial Altruism in Humans." *Nature* 442:912–915.
Bijma, P., and M. Wade. (2008). "The Joint Effects of Kin, Multilevel Selection and

Indirect Genetic Effects on Response to Genetic Selection." *Journal of Evolutionary Biology* 21 (5): 1178–1188.
Boehm, C. (1999). *Hierarchy in the Forest: The Evolution of Egalitarian Behavior.* Cambridge, MA: Harvard University Press.
Boelen, B. J. (1959). *Symposium on Evolution, Held in Commemoration of the Centenary of Charles Darwin's "The Origin of Species."* Pittsburgh, PA: Duquesne University Press.
Borrello, M. E. (2003). "Synthesis and Selection: Wynne-Edwards' Challenge to David Lack." *Journal of the History of Biology* 36 (3): 531–566.
———. (2004). "Mutual Aid and Animal Dispersion: An Historical Analysis of Alternatives to Darwin." *Perspectives in Biology and Medicine* 47 (1): 15–31.
———. (2005). "The Rise, Fall and Resurrection of Group Selection." *Endeavour* 29 (1): 43–47.
Bowler, P. J. (1983). *The Eclipse of Darwinism: Anti-Darwinian Evolution Theories in the Decades around 1900.* Baltimore, MD: Johns Hopkins University Press.
———. (1988). *The Non-Darwinian Revolution: Reinterpreting a Historical Myth.* Baltimore, MD: Johns Hopkins University Press.
———. (1989). *Evolution: The History of an Idea.* Berkeley, CA: University of California Press.
———. (1993). *Biology and Social Thought: 1850–1914.* Berkeley: Regents of the University of California.
Bowles, S., and H. Gintis. (2002). "Social Capital and Community Governance." *Economic Journal* 112 (483): F419–F436.
Boyd, R., H. Gintis, et al. (2003). "The Evolution of Altruistic Punishment." *Proceedings of the National Academy of Sciences* 100 (6): 3531–3535.
Braestrup, F. W. (1963). "Special Review." *Oikos* 14 (1): 113–120.
Brandon, R. N. (1990). *Adaptation and Environment.* Princeton, NJ: Princeton University Press.
———. (1996). *Concepts and Methods in Evolutionary Biology.* Cambridge: Cambridge University Press.
Brandon, R. N., and R. Burian, eds. (1984). *Genes, Organisms, Populations: Controversies over the Units of Selection.* Cambridge, MA: MIT Press.
Brooke, J. H. (1991). *Science and Religion: Some Historical Perspectives.* Cambridge: Cambridge University Press.
Burkhardt, R. (1981). "On the Emergence of Ethology as a Scientific Discipline." *Conspectus of History* 1 (7): 62–81.
———. (1983). "The Development of an Evolutionary Ethology." In *Evolution from Molecules to Men,* ed. D. Bendall, 429–444. Cambridge: Cambridge University Press.
———. (1985). "Darwin on Animal Behavior and Evolution." In *The Darwinian Heritage,* ed. D. Kohn, 327–365. Princeton, NJ: Princeton University Press.
———. (1992). "Huxley and the Rise of Ethology." In *Julian Huxley: Biologist and Statesman of Science,* ed. C. K. Waters and A. V. Helden, 127–149. Houston, TX: Rice University Press.
———. (1999). "Ethology, Natural History, the Life Sciences, and the Problem of Place." *Journal of the History of Biology* 32:489–509.

———. (2005). *Patterns of Behavior: Konrad Lorenz, Niko Tinbergen and the Founding of Ethology*. Chicago: University of Chicago Press.

Buss, L. W. (1983). "Evolution, Development, and the Units of Selection." *Proceedings of the National Academy of Sciences, USA* 80 (March): 1387–1391.

———. (1987). *The Evolution of Individuality*. Princeton, NJ: Princeton University Press.

Cain, J., and M. Ruse, eds. (2009). *Descended from Darwin: Insights into the History of Evolutionary Studies, 1900–1970*. Philadelphia: American Philosophical Society Press.

Caplan, A. L., ed. (1978). *The Sociobiology Debate: Readings on Ethical and Scientific Issues*. New York: Harper and Row.

Carpenter, G. D. H., and E. B. Ford. (1933). *Mimicry*. London: Methuen.

Chitty, D. (1996). *Do Lemmings Commit Suicide? Beautiful Hypotheses and Ugly Facts*. New York: Oxford University Press.

Churchill, F. B. (1970). "The History of Embryology as Intellectual History." *Journal of the History of Biology* 3 (1): 155–181.

———. (1977). "The Weismann-Spencer Controversy over the Inheritance of Acquired Characters." In *Human Implications of Scientific Advance*. Edinburgh: Edinburgh University Press.

———. (1981). "In Search of the New Biology: An Epilogue." *Journal of the History of Biology* 12 (1): 177–191.

Cope, E. D. (1893). "Heredity in the Social Colonies of the Hymenoptera." *Proceedings of the Academy of Natural Sciences of Philadelphia*, 436–438.

Corning, P. (2005). *Holistic Darwinism: Synergy, Cybernetics and the Bioeconomics of Evolution*. Chicago: University of Chicago Press.

Cort, D. (1963). "The Glossy Rats: A Review of *Animal Dispersion in relation to Social Behavior*." *Nation* 197 (November 16): 327.

Cronin, H. (1991). *The Ant and the Peacock: Altruism and Sexual Selection from Darwin to Today*. Cambridge: Cambridge University Press.

Crowcroft, P. (1991). *Elton's Ecologists: A History of the Bureau of Animal Population*. Chicago: University of Chicago Press.

Cunningham, J. T. (1894). "Neuter Insects and Darwinism." *Natural Science*, April 1894, 281–289.

———. (1898). "The Species, the Sex and the Individual." *Natural Science*, September 1898, 184–192, 233–239.

———. (1900). *Sexual Dimorphism in the Animal Kingdom: A Theory of the Evolution of Secondary Sexual Characters*. London: Adam and Charles Black.

Darwin, C. (1859). *On the Origin of Species*. Cambridge, MA: Harvard University Press.

———. (1958). *The Autobiography of Charles Darwin, 1809–1882*. New York: W. W. Norton.

———. (1981). *The Descent of Man, and Selection in relation to Sex*. Princeton, NJ: Princeton University Press. Originally published 1871.

Daston, L., and F. Vidal, eds. (2004). *The Moral Authority of Nature*. Chicago: University of Chicago Press.

Dawkins, R. (1976). *The Selfish Gene*. Oxford: Oxford University Press.

———. (1989). *The Selfish Gene,* new ed. London: Oxford University Press.

Depew, D., and B. H. Weber, eds. (1985). *Evolution at a Crossroads: The New Biology and the New Philosophy of Science.* Cambridge: MIT Press.

Dewsbury, D. A., ed. (1985). *Leaders in the Study of Animal Behavior: Autobiographical Perspectives.* Lewisburg, PA: Bucknell University Press.

Dietrich, M. R. (1998). "Paradox and Persuasion: Negotiating the Place of Molecular Evolution within Evolutionary Biology." *Journal of the History of Biology* 31:85–111.

Dixon, T. (2008). *The Invention of Altruism: Making Moral Meanings in Victorian Britain.* Oxford: Oxford University Press.

Dobzhansky, T. (1937). *Genetics and the Origin of Species.* Ed. Leslie C. Dunn. New York: Columbia University Press.

———. (1951). "Mendelian Populations and Their Evolution." in *Genetics in the Twentieth Century,* ed. L. C. Dunn, 573–589. New York: Macmillan.

———. (1966). "Are Naturalists Old-Fashioned?" *American Naturalist* 100 (915): 541–550.

———. (1970). *The Genetics of the Evolutionary Process.* New York: Columbia University Press.

———. (1973). "Nothing in Biology Makes Sense Except in the Light of Evolution." *American Biology Teacher* 35:125–129.

Dugatkin, L. (1999). *Cheating Monkeys and Citizen Bees: The Nature of Cooperation in Animals and Humans.* New York: Free Press.

———. (2006). *The Altruism Equation: Seven Scientists Search for the Origins of Goodness.* Princeton, NJ: Princeton University Press.

Dugatkin, L. A., and H. K. Reeve. (1994). "Behavioral Ecology and Levels of Selection: Dissolving the Group Selection Controversy." *Advances in the Study of Behavior* 23:101–133.

Ebling, F. J., and D. M. Stoddart, eds. (1978). *Population Control by Social Behaviour.* Symposia of the Institute of Biology. London: Institute of Biology.

Elton, C. (1963). "Self-Regulation of Animal Populations." *Nature* 197 (February 16): 634.

Gayon, J. (1998). *Darwinism's Struggle for Survival: Heredity and the Hypothesis of Natural Selection.* Trans. Matthew Cobb. Cambridge: Cambridge University Press.

Gintis, H., S. Bowles, et al., eds. (2005). *Moral Sentiments and Material Interests: The Foundations of Cooperation in Economic Life.* Economic Learning and Social Evolution. Cambridge, MA: MIT Press.

Glick, T. F., ed. (1988). *The Comparative Reception of Darwinism.* Chicago: University of Chicago Press.

Goldsmith, E., et al. (1972). *Blueprint for Survival.* Boston: Houghton Mifflin.

Gould, S. J. (1983). "The Hardening of the Modern Synthesis." In *Dimensions of Darwinism: Themes and Counterthemes in Twentieth Century Evolutionary Theory,* ed. M. Grene, 71–93. Cambridge: Cambridge University Press.

———. (1991). "Kropotkin Was No Crackpot." In *Bully for Brontosaurus,* 325–339. New York: W. W. Norton.

———. (2002). *The Structure of Evolutionary Theory*. Cambridge, MA: Harvard University Press.

Gould, S. J., and R. Lewontin. (1978). "The Spandrels of San Marco and the Panglossian Paradigm." *Proceedings of the Royal Society of London* 205:581–598.

Grant, P., and R. Grant. (2007). *How and Why Species Multiply: The Radiation of Darwin's Finches*. Princeton, NJ: Princeton University Press.

Greene, J. (1977). "Darwin as Social Evolutionist." *Journal of the History of Biology* 10:1–27.

Grene, M., ed. (1983). *Dimensions of Darwinism: Themes and Counterthemes in Twentieth-Century Evolutionary Theory*. Cambridge: Cambridge University Press.

Griesemer, J. R., and M. J. Wade. (1988). "Laboratory Models, Causal Explanation, and Group Selection." *Biology and Philosophy* 3:67–96.

Hagen, J. (1992). *An Entangled Bank: The Origins of Ecosystem Ecology*. New Brunswick, NJ: Rutgers University Press.

Haldane, J. B. S. (1932). *The Causes of Evolution*. Ithaca, NY: Cornell University Press.

Hamilton, W. D. (1963). "The Evolution of Altruistic Behavior." *American Naturalist* 97:354–356.

———. (1964). "The Genetical Evolution of Social Behavior, I and II." *Journal of Theoretical Biology* 7:1–52.

———. (1996). *Narrow Roads of Gene Land: The Collected Papers of W. D. Hamilton*. Vol. 1. *Evolution of Social Behavior.* Oxford: W. H. Freeman.

———. (1970). "Selfish and Spiteful Behavior in an Evolutionary Model." *Nature* 228 (5277): 1218–1220.

Hanson, N. R. (1988). "Observation." In *Introductory Readings in the Philosophy of Science*, ed. E. D. Klemke, R. Hollinger, and A. D. Kline, 184–195. Buffalo, NY: Prometheus Books.

Harman, O., and M. R. Dietrich, eds. (2008). *Rebels, Mavericks, and Heretics in Biology*. New Haven, CT: Yale University Press.

Hinde, R. A. (1985). "Ethology in relation to Other Disciplines." In *Leaders in the Study of Animal Behavior: Autobiographical Perspectives,* ed. D. Dewsbury. Lewisburg, PA: Bucknell University Pres.

Hofstadter, R. (1955). *Social Darwinism in American Thought*. Boston: Beacon Press.

Hölldobler, B., and E. O. Wilson. (2008). *The Superorganism: The Beauty, Elegance, and Strangeness of Insect Societies*. New York: W. W. Norton.

Hull, D. L. (1976). "Are Species Really Individuals?" *Systematic Zoology* 25: 174–191.

———. (1980). "Individuality and Selection." *Annual Review of Ecology and Systematics* 11:311–332.

———. (1988). *Science as a Process: An Evolutionary Account of the Social and Conceptual Development of Science*. Chicago: University of Chicago Press.

Huxley, J. S. (1912). *The Individual in the Animal Kingdom*. Cambridge: Cambridge University Press.

———. (1942). *Evolution: The Modern Synthesis.* London: Allen and Unwin.
———. (1963). "Lorenzian Ethology." *Zeitschrift für Tierpsychologie* 20:407.
Huxley, T. H. (1893). "The Struggle for Existence in Human Society." In *Evolution and Ethics,* by T. H. Huxley, 9:195–236. London: Macmillan..
Jenkins, D., and A. Watson. (1997). "Obituary: Vero Copner Wynne-Edwards (1906–1997)." *Ibis* 139 (2): 415–418.
Jordan, D. S., and V. L. Kellogg. (1907). *Evolution and Animal Life: An Elementary Discussion of Facts, Processes, Laws and Theories relating to the Life and Evolution of Animals.* New York: D. Appleton.
Kaye, H. L. (1986). *The Social Meaning of Modern Biology: From Social Darwinism to Sociobiology*. New Haven, CT: Yale University Press.
Keller, E. F., and E. A. Lloyd, eds. (1992). *Keywords in Evolutionary Biology*. Cambridge, MA: Harvard University Press.
Kellogg, V. L. (1908). *Darwinism To-day: A Discussion of Present-Day Criticism of the Darwinian Selection Theories, Together with a Brief Account of the Principal Other Proposed Auxiliary and Alternative Theories of Species-Forming*. New York: Henry Holt.
Kerr, B., C. Neuhauser, et al. (2006). "Local Migration Promotes Competitive Restraint in a Host-Pathogen 'Tragedy of the Commons.'" *Nature* 442:75–78.
King, J. A. (1965). "Social Behavior and Population Homeostasis." *Ecology* 46:210.
Kingsland, S. E. (1985). *Modeling Nature: Episodes in the History of Population Ecology*. Chicago: University of Chicago Press.
Kitcher, P. (1985). *Vaulting Ambition: Sociobiology and the Quest for Human Nature*. Cambridge: MIT press.
Klemke, E. D., R. Hollinger, et al., eds. (1988). *Introductory Readings in the Philosophy of Science*. Buffalo, NY: Prometheus Books.
Klopfer, P. H. (1999). *Politics and People in Ethology.* London: Associated University Press.
Kohler, R. E. (1982). *From Medical Chemistry to Biochemistry: The Making of a Biomedical Discipline*. Cambridge: Cambridge University Press.
———. (1994). *Lords of the Fly:* Drosophila *Genetics and the Experimental Life.* Chicago: University of Chicago Press.
Krebs, J. R. (1992). "The Book That Most . . ." *Biologist* 39:217.
Kropotkin, P. A. (1902). *Mutual Aid: A Factor in Evolution*. New York: McClure Philips.
———. (1910). "The Direct Action of Environment on Plants." *Nineteenth Century and After* 68 (January): 58–77.
———. (1910). "The Response of Animals to Their Environment." *Nineteenth Century and After* 68 (November): 856–867, 1047–1059.
———. (1910). "The Theory of Evolution and Mutual Aid." *Nineteenth Century and After* 67 (January): 86–107.
———. (1912). "Inheritance of Acquired Characters: Theoretical Difficulties." *Nineteenth Century and After* 71 (March): 511–531.
———. (1915). "Inherited Variation in Animals." *Nineteenth Century and After* 78 (November): 1124–1144.

———. (1919). "The Direct Action of the Environment and Evolution." *Nineteenth Century and After* 85 (January): 70–89.

———. (1955). *Mutual Aid: A Factor of Evolution*. Boston: Extending Horizons Books.

Lack, D. (1952). "Reproductive Rate and Population Density in the Great Tit: Kluijver's Study." *Ibis* 94:166–173.

———. (1954). "The Evolution of Reproductive Rates." In *Evolution as a Process*, ed. J. S. Huxley, A. C. Hardy, and E. B. Ford, 143–156. London: Allen and Unwin.

———. (1954). *The Natural Regulation of Animal Numbers*. Oxford: Clarendon.

———. (1954). "The Stability of the Heron Population." *British Birds* 47:111–119.

———. (1966). *Population Studies of Birds*. Oxford: Clarendon.

———. (1973). "My Life as an Amateur Ornithologist." *Ibis* 115:421–431.

Leeper, G. W. (1962). *The Evolution of Living Organisms; A Symposium to Mark the Centenary of Darwin's "Origin of Species" and of the Royal Society of Victoria,* Melbourne, December 1959. Parkville: Melbourne University Press.

Lewontin, R. C. (1970). "The Units of Selection." *Annual Review of Ecology and Systematics* 1:1–18.

———. (1978). "Adaptation." *Scientific American* 239:156–169.

Lewontin, R. C., S. Rose, and L. J. Kamin. (1984). *Not in Our Genes: Biology, Ideology and Human Nature*. New York: Pantheon Books.

Lloyd, E. A. (1988). *The Structure and Confirmation of Evolutionary Theory*. Princeton, NJ: Princeton University Press.

———. (1999). "Altruism Revisited." *Quarterly Review of Biology* 74:447–449.

———. (2001). "Units and Levels of Selection: An Anatomy of the Units of Selection Debates." In *Thinking about Evolution: Historical, Philosophical and Political Perspectives,* ed. R. S. Singh, C. B. Krimbas, D. B. Paul, and J. Beatty, 267–291. Cambridge: Cambridge University Press.

Lloyd, E., and S. J. Gould. (1993). "Species Selection on Variability." *Proceedings of the National Academy of Sciences* 90:595–599.

Lorenz, K. (1970). *Studies in Animal and Human Behavior.* Cambridge, MA: Harvard University Press.

———. (1971). "Part and Parcel in Animal and Human Societies: A Methodological Discussion." In *Studies in Animal and Human Behavior,* ed. and trans. R. Martin, 2:115–195. Cambridge, MA: Harvard University Press.

———. (1974). "Analogy as a Source of Knowledge." *Science* 185:229–234.

———. (1987). *The Waning of Humaneness*. Boston: Little, Brown.

———. (1990). *On Life and Living: Konrad Lorenz in Conversation with Kurt Mündl.* New York: St. Martin's Press.

Maynard Smith, J. (1964). "Group Selection and Kin Selection." *Nature* 201: 1145–1147.

———. (1976). "Group Selection." *Quarterly Review of Biology* 51:277–283.

Mayr, E. (1942). *Systematics and the Origin of Species*. New York: Columbia University Press.

———. (1963). *Animal Species and Evolution*. Cambridge, MA: Harvard University Press.

———. (1963). "The New versus the Classical in Science." *Science* 141 (3583): 763.

———. (1993). "What Was the Evolutionary Synthesis?" *Trends in Ecology and Evolution* 8:31–34.

Mayr, E., and W. B. Provine, eds. (1980). *The Evolutionary Synthesis: Perspectives on the Unification of Biology*. Cambridge, MA: Harvard University Press.

McIntosh, R. P. (1989). "Citation Classics of Ecology." *Quarterly Review of Biology* 64 (1): 31–49.

Mitchell, S. D. (1995). "The Superorganism Metaphor: Then and Now." In *Biology as Society, Society as Biology: Metaphors,* ed. S. Massen, E. Mendelsohn, and P. Weingart, 231–247. Dordrecht: Kluwer Academic Publishers.

Mitman, G. (1988). "From Population to Society: The Cooperative Metaphors of W. C. Allee and A. E. Emerson." *Journal of the History of Biology* 21 (2): 173–194.

———. (1990). "Evolution as Gospel: William Patten, the Language of Democracy, and the Great War." *Isis* 81:446–463.

———. (1992). *The State of Nature: Ecology, Community and American Social Thought, 1900–1950*. Chicago: University of Chicago Press.

Mogie, Michael. (1996). "Malthus and Darwin: World Views Apart." *Evolution* 50:2086–2088.

Morgan, T. H. (1903). *Evolution and Adaptation.* New York: Macmillan.

———. (1932). *The Scientific Basis of Evolution*. New York: W. W. Norton.

Morrell, J. (1994). "The Non-medical Sciences, 1914–1939." In *The History of the University of Oxford,* ed. B. Harrison, 8:139–163. Oxford: Clarendon Press.

———. (1997). *Science at Oxford, 1914–1939: Transforming an Arts University*. Oxford: Clarendon Press.

"The Nature of Social Life." (1962). *Times Literary Supplement,* December 14, 967.

Newton, I. "Vero Copner Wynne-Edwards C.B.E." Biographical Memoirs of the Fellows of the Royal Society 44 (1998): 473–484.

Nitecki, M., ed. (1988). *Evolutionary Progress*. Chicago: University of Chicago Press.

Okasha, S. (2006). *Evolution and the Levels of Selection*. New York: Oxford University Press.

Papworth, J. (1991). "Mass Society and the Pheasant." *Ecologist* 21 (5): 232.

Park, T. (1961). "An Ecologist's View." *Bulletin of the Ecological Society of America* 42:4–10.

Pauly, P. J. (1987). *Controlling Life: Jacques Loeb and the Engineering Ideal in Biology*. New York: Oxford University Press.

Perrins, C. M. (1963). "Survival in the Great Tit, *Parus major*." *Proceedings of the International Ornithological Congress* 13:717–728.

———. (1965). "Population Fluctuations and Clutch-Size in the Great Tit *Parus major*." *Journal of Animal Ecology* 34:601–647.

Pollock, G. B. (1989). "Suspending Disbelief—of Wynne-Edwards and His Reception." *Journal of Evolutionary Biology* 2:205–221.

Provine, W. B. (1971). *The Origins of Theoretical Population Genetics*. Chicago: University of Chicago Press.

Rainger, R., K. R. Benson, et al., eds. (1988). *The American Development of Biology*. Philadelphia: University of Pennsylvania Press.
Richards, R. J. (1977). "Lloyd Morgan's Theory on Instinct: From Darwinism to Neo-Darwinism." *Journal of the History of the Behavioral Sciences* 13:12–32.
———. (1982). "The Emergence of Evolutionary Biology of Behavior in the Early 19th Century." *British Journal of the History of Science* 15:241–280.
———. (1987). *Darwin and the Emergence of Evolutionary Theories of Mind and Behavior*. Chicago: University of Chicago Press.
———. (1992). *The Meaning of Evolution: The Morphological Construction and Ideological Reconstruction of Darwin's Theory*. Chicago: University of Chicago Press.
Richerson, P., and R. Boyd. (2005). *Not by Genes Alone: How Culture Transformed Human Evolution*. Chicago: University of Chicago Press.
Ridley, M. (1982). "Coadaptation and the Inadequacy of Natural Selection." *British Journal of the History of Science* 15:45–68.
———. (1987). "*Evolution through Group Selection*—Review." *Ethology* 74 (3): 260–261.
Riley, C. V. (1894). "On Social Insects and Evolution." *Report of the British Association for the Advancement of Science*, 689–691.
———. (1894). "Social Insects from Psychical and Evolutional Points of View." *Proceedings of the Biological Society of Washington* 9 (April): 1–74.
Ruse, M. (1980). "Charles Darwin and Group Selection." *Annals of Science* 37:615–630.
———. (1980). "Social Darwinism: The Two Sources." *Albion* 12:23–36.
Sapp, J. (1994). *Evolution by Association: A History of Symbiosis*. New York: Oxford University Press.
Schwartz, J. S. (1999). "Robert Chambers and Thomas Henry Huxley, Science Correspondents: The Popularization and Dissemination of Nineteenth Century Natural Science." *Journal of the History of Biology* 32:343–383.
Segerstråle, U. (2000). *Defenders of the Truth: The Battle for Science in the Sociobiology Debate and Beyond*. Oxford: Oxford University Press.
Shavit, A. (2004). "Shifting Values Partly Explain the Debate over Group Selection." *Studies in History and Philosophy of Biological and Biomedical Sciences* 35 (4): 697–720.
Simpson, G. G. (1944). *Tempo and Mode in Evolution*. New York: Columbia University Press.
Simpson, J. A., and E. S. C. Weiner, eds. (1991). *The Oxford English Dictionary*. Oxford: Clarendon Press.
Sinclair, A. R. E. (1989). "Population Regulation in Animals." *British Ecological Society Symposium* 29:197–241.
Sleigh, C. (2007). *Six Legs Better: A Cultural History of Myrmecology*. Baltimore, MD: Johns Hopkins University Press.
Smocovitis, V. B. (1996). *Unifying Biology: The Evolutionary Synthesis and Evolutionary Biology*. Princeton, NJ: Princeton University Press.
Snow, D. W. (1958). "The Breeding of the Blackbird, *Turdus merula*, at Oxford." *Ibis* 100:1–30.

Sober, E. (1984). *The Nature of Selection: Evolutionary Theory in Philosophical Focus*. Chicago: University of Chicago Press.
Sober, E., and R. C. Lewontin. (1982). "Artifact, Cause, and Genic Selection." *Philosophy of Science* 49:157–180.
Sober, E., and D. S. Wilson. (1998). *Unto Others: The Evolution and Psychology of Unselfish Behavior*. Cambridge, MA: Harvard University Press.
Southern, H. N. (1959). "Mortality and Population Control." *Ibis* 101:429–436.
Spencer, H. (1866). *The Principles of Biology*. New York: D. Appleton.
Sterelny, K. (1996). "The Return of the Group." *Philosophy of Science* 63 (December): 562–584.
Sterelny, K., and P. Kitcher. (1988). "The Return of the Gene." *Journal of Philosophy* 85:339–361.
Tax, S., ed. (1960). *Evolution after Darwin.* University of Chicago Centennial. Chicago: University of Chicago Press.
Thomson, J. A. (1896). "The Endeavour after Well-Being." *Natural Science,* January,) 21–26.
———. (1925). *Concerning Evolution*. New Haven, CT: Yale University Press.
Tinbergen, N. (1965). "Behavior and Natural Selection." In *Ideas in Modern Biology,* ed. J. Moore. Garden City, NJ: Natural History Press.
Todes, D. P. (1987). "Darwin's Malthusian Metaphor and Russian Evolutionary Thought, 1859–1917." *Isis* 78:537–551.
———. (1989). *Darwin without Malthus: The Struggle for Existence in Russian Evolutionary Thought.* New York: Oxford University Press.
Trivers, R. (1971). "The Evolution of Reciprocal Altruism." *Quarterly Review of Biology* 46:33–57.
Wade, M. J. (1976). "Group Selection among Laboratory Populations of *Tribolium*." *Proceedings of the National Academy of Sciences* 73:4604–4607.
———. (1977). "Experimental Study of Group Selection." *Evolution* 31:134–153.
———. (1978). "A Critical Review of the Models of Group Selection." *Quarterly Review of Biology* 53:101–114.
———. (1980). "An Experimental Study of Kin Selection." *Evolution* 34 (5): 844–855.
———. (1980). "Kin Selection: Its Components." *Science* 210:665–667.
———. (1984). "The Changes in Group Selected Traits When Group Selection Is Relaxed." *Evolution* 38:1039–1046.
———. (1985). "Soft Selection, Hard Selection, Kin Selection and Group Selection." *American Naturalist* 125:61–73.
———. (1988). "*Evolution through Group Selection*." *Evolution* 42 (5): 1116.
Wallace, A. R. (1889). *Darwinism*. London: Macmillan.
Wanstall, P. J., ed. (1961). *A Darwin Centenary: The Report of the Conference Held by the Botanical Society of the British Isles in 1959 to Mark the Centenary of the Publication of "The Origin of Species."* London: Botanical Society of the British Isles.
Waters, C. M., and B. L. Bassler. (2005). "Quorum Sensing: Cell to Cell Communication in Bacteria." *Annual Review of Cell and Developmental Biology* 21:319–346.

Watson, J. D., and F. Crick. (1953). "A Structure for Deoxyribose Nucleic Acid." *Nature* 171:737–738.

Weiner, J. (1994). *The Beak of the Finch: A Story of Evolution in Our Time*. New York: Knopf.

Weismann, A. (1889). *Essays upon Heredity and Kindred Biological Problems*. Oxford: Clarendon Press.

———. (1893). *The Germ-Plasm: A Theory of Heredity.* New York: C. Scribner's Sons.

———. (1904). *The Evolution Theory*. London: Edward Arnold.

Werfel, J., and Y. Bar-Yam. (2004). "The Evolution of Reproductive Restraint through Social Communication." *Proceedings of the National Academy of Sciences* 101 (30): 11019–11024.

West, S. A., A. S. Griffin, and A. Gardener. (2007). "Social Semantics: Altruism, Cooperation, Mutualism, Strong Reciprocity, and Group Selection." *Journal of Evolutionary Biology* 20:415–432.

West-Eberhard, M. J. (1986). "*Evolution through Group Selection*—Review." *Journal of Genetics* 65:213–217.

Wheeler, W. M. (1910). *Ants: Their Structure, Development and Behavior*. New York: Norton.

———. (1911). "The Ant-Colony as an Organism." *Journal of Morphology* 22:307–325.

———. (1928). *Emergent Evolution and the Development of Societies*. New York: Norton.

Wiens, J. A. (1966). "On Group Selection and Wynne-Edwards' Hypothesis." *American Scientist* 54 (3): 273–287.

Williams, G. C. (1966). *Adaptation and Natural Selection: A Critique of Some Current Evolutionary Thought*. Princeton, NJ: Princeton University Press.

———. (1996). *Adaptation and Natural Selection: A Critique of Some Current Evolutionary Thought,* rev. ed. Princeton, NJ: Princeton University Press.

———, ed. (1971). *Group Selection*. Chicago: Aldine Atherton.

———. (1992). *Natural Selection: Domains, Levels and Challenges*. New York: Oxford University Press.

Williams, G. C., and D. C. Williams. (1957). "Natural Selection of Individually Harmful Social Adaptations among Sibs, with Special Reference to Social Insects." *Evolution* 11 (March): 32–39.

Wilson, D. S. (1975). "A Theory of Group Selection." *Proceedings of the National Academy of Science* 72:143–146.

———. (1976). "Evolution on the Level of Communities." *Science* 192:1358–1360.

———. (1977). "Structured Demes and the Evolution of Group-Advantageous Traits." *American Naturalist* 111 (977): 157–185.

———. (1980). *The Natural Selection of Populations and Communities*. Menlo Park, CA: Benjamin/Cummings.

———. (1983). "The Group Selection Controversy: History and Current Status." *Annual Review of Ecology and Systematics* 14:159–187.

———. (2007). *Evolution for Everyone: How Darwin's Theory Can Change the Way We Think about Our Lives*. New York: Delacorte Press.

———. (2008). "Social Semantics: Toward a Genuine Pluralism in the Study of Social Behavior." *Journal of Evolutionary Biology* 21:368–373.

Wilson, D. S., and E. Sober. (1989). "Reviving the Superorganism." *Journal of Theoretical Biology* 136:337–356.

———. (1994). "Reintroducing Group Selection to the Human Sciences." *Behavioral and Brain Sciences* 17 (4): 585–654.

Wilson, D. S., and E. O. Wilson. (2007). "Rethinking the Theoretical Foundations of Sociobiology." *Quarterly Review of Biology* 82 (4): 327–348.

Wilson, E. O. (1968). "The Superorganism Concept and Beyond." *Effet de Groupe Chez les Animaux: Colloques Internationaux du Centre National de Recherche Scientifique* 173:27–39.

———. (1973). "Group Selection and Its Significance for Ecology." *BioScience* 23 (11): 631–638.

———. (1975). *Sociobiology: The New Synthesis*. Cambridge, MA: Belknap Press of Harvard University Press.

———. (1994). *Naturalist*. Washington, DC: Island Press.

Wilson, E. O., and B. Hölldobler. (2005). "Eusociality: Origin and Consequences." *Proceedings of the National Academy of Sciences* 102 (38): 13367–13371.

Wimsatt, W. (1980). "Reductionist Research Strategies and Their Biases in the Units of Selection Controversy." In *Scientific Discovery: Case Studies,* ed. T. Nickles, 213–259. Dordrecht: Reidel.

———. (1981). "Units of Selection and the Structure of the Multi-level Genome." *Proceedings of the Philosophy of Science Association* 2:122–183.

Wright, S. (1929). "Fisher's Theory of Dominance." *American Naturalist* 63:274–279.

———. (1931). "Evolution in Mendelian Populations." *Genetics* 16:97–159.

———. (1932). *The Roles of Mutation, Inbreeding, Crossbreeding and Selection in Evolution.* Ithaca, NY: Cornell University Press.

———. (1938). "Size of Population and Breeding Structure in relation to Evolution." *Science* 87:430–431.

———. (1945). "Tempo and Mode in Evolution: A Critical Review." *Ecology* 26:415–419.

———. (1969). *Evolution and the Genetics of Populations.* Vol. 2. *Theory of Gene Frequencies*. Chicago: University of Chicago Press.

———. (1980). "Genic and Organismic Selection." *Evolution* 34 (5): 825–843.

Wynne-Edwards, V. C. (1929). "The Behaviour of Starlings in Winter: An Investigation of the Diurnal Movements and Social Roosting-Habit." *British Birds* 23:138–153, 170–180.

———. (1929). "On the Waking-Time of the Nightjar." *Journal of Experimental Biology* 7:241–247.

———. (1931). "The Behaviour of Starlings in Winter." *British Birds* 24:346–353.

———. (1935). "On the Habits and Distribution of Birds on the North Atlantic." *Proceedings of the Boston Society of Natural History* 40 (4): 233–346.

———. (1939). "Intermittent Breeding of the Fulmar (*Fulmarus glacialis*), with Some General Observations on Non-breeding in Sea-Birds." *Proceedings of the Zoological Society of London,* 127–132.

———. (1948). "The Nature of Subspecies." *Scottish Naturalist* 60:195–208.

———. (1952). "Geographical Variation in the Bill of the Fulmar (*Fulmaris glacialis*)." *Scottish Naturalist* 64:84–101.

———. (1952). "Zoology of the Baird Expedition." *Auk* 69 (4): 353–391.

———. (1955). "Low Reproductive Rates in Birds, Especially Sea-Birds." *Acta of the Eleventh International Congress of Ornithology,* 540–547.

———. (1959). "The Control of Population Density through Social Behaviour: A Hypothesis." *Ibis* 101:436–441.

———. (1962). *Animal Dispersion in relation to Social Behavior.* Edinburgh: Oliver and Boyd.

———. (1963). "Intergroup Selection in the Evolution of Social Systems." *Nature* 200 (4907): 623–626.

———. (1964). "Group Selection and Kin Selection: Response to Maynard-Smith." *Nature* 201 (4924): 1145–1147.

———. (1965). "Self-Regulating Systems in Populations of Animals: A New Hypothesis Illuminates Aspects of Animal Behavior That Have Hitherto Seemed Unexplainable." *Science* 147 (March 26, 1965): 1543–1548.

———. (1971). "Space Use and the Social Community in Animals and Men." In *Behaviour and Environment: The Use of Space in Animals and Men,* ed. A. H. Esser, 267–280. New York: Plenum Press.

———. (1972). "Ecology and the Evolution of Social Ethics." In *Biology and the Human Sciences: The Herbert Spencer Lectures, 1970,* ed. J. W. S. Pringle, 49–69. Oxford: Oxford University Press.

———. (1977). "Society versus the Individual in Animal Evolution." In *Evolutionary Ecology,* ed. B. Stonehouse and C. M. Perrins, 5–18. Baltimore, MD: University Park Press.

———. (1978). "Intrinsic Population Control: An Introduction." In *Population Control by Social Behavior,* ed. F. J. Ebling and D. M. Stoddart, 1–22. London: Institute of Biology.

———. (1982). "*The Genetics of Altruism*—Review." *Social Science and Medicine* 16:1096.

———. (1985). "Backstage and Upstage with *Animal Dispersion*." In *Leaders in the Study of Animal Behavior: Autobiographical Perspectives,* ed. D. Dewsbury, 486–512. Lewisburg, PA: Bucknell University Press.

———. (1986). *Evolution through Group Selection.* Oxford: Blackwell Scientific Publications.

———. (1991). "Ecology Denies Neo-Darwinism." *Ecologist* 21 (3): 136–141.

———. (1993). "A Rationale for Group Selection." *Journal of Theoretical Biology* 162:1–22.

Wynne-Edwards, V. C., R. M. Lockley, et al. (1936). "The Distribution and Number of Breeding Gannets." *British Birds* 29:262–276.

Index